INTERNET ALLEY

LEMELSON CENTER STUDIES IN INVENTION AND
INNOVATION
ARTHUR P. MOLELLA AND JOYCE BEDI, GENERAL EDITORS

Arthur P. Molella and Joyce Bedi, editors, *Inventing for the Environment*

Paul E. Ceruzzi, *Internet Alley: High Technology in Tysons Corner, 1945–2005*

INTERNET ALLEY

HIGH TECHNOLOGY IN TYSONS CORNER, 1945–2005

PAUL E. CERUZZI

THE MIT PRESS

CAMBRIDGE, MASSACHUSETTS

LONDON, ENGLAND

For information about special quantity discounts, please e-mail special_sales@
mitpress.mit.edu

This book was set in Engravers Gothic and Bembo by SPi, Pondicherry, India.
Printed and bound in the United States of America.

Library of Congress Cataloging-in-Publication Data

Ceruzzi, Paul E.
Internet alley: high technology in Tysons Corner, 1945–2005 / By Paul E. Ceruzzi.
     p. cm.
Includes bibliographical references and index.
ISBN 978-0-262-03374-9 (hardcover : alk. paper)
1. Internet–History. I. Title.
Gift 4/09
TK5105.875.I57C437 2007
004.67'8—dc22

200701893

10 9 8 7 6 5 4 3 2 1

CONTENTS

Not long after I moved to the Washington, D.C., area, I found myself driving through a place known as Tysons Corner, Virginia, near the intersection of Routes 7, 123, and the Washington Beltway. I could not help but notice the sleek, new office buildings clustered next to one another in what was nominally a residential suburb. I had seen such buildings before, but not such a concentration, and not in a place that I assumed was far from the centers of political, technical, or financial power in the Washington region.

Many of the buildings had the names of their tenants displayed in bold lettering on the top floor. Some names suggested high technology companies: names ending in "-tronics," "-ex," or the like. Others consisted of three-letter acronyms, few of which I recognized. As I drove by, all I could think of was the famous line from *Butch Cassidy and the Sundance Kid*: "Who *are* those guys"?

The name "Tysons Corner" was not unknown to me. Before moving to Washington, I had read in a newsletter for amateur radio operators that it was the probable source of the mysterious 'numbers' stations—the top-secret broadcasts of random sets of numbers heard on high-frequency radio bands whose purpose and origin were a mystery. The very next issue of that newsletter issued a retraction: Tysons Corner was not the source of these transmissions after all. That only made it worse. Nothing quickens the pulse of a conspiracy theorist as much as a semi-official denial. As I drove through the thicket of new office buildings I noticed, across from the Tysons Galleria shopping mall, a government-owned radio tower. A sign at the base warned visitors not to photograph or make sketches of the tower, citing the Internal Security Act of 1950. It seemed odd to find this government facility next to an upscale shopping mall. This only deepened the mystique. Who *are* those guys?

FIGURE P. I

Tysons Corner, circa 1989 looking west along Route 7, or Leesburg Pike. Route 123
crosses on a overpass at the top of the photo. The original "corner" formed by that
intersection gave the region its name. The radio tower, now surrounded and obscured
by high-rise office buildings, is visible to the right of the overpass. The modest, cube-
shaped building to the left of the highway, center, was the first Tysons location for
BDM, discussed in chapter 5. The building was torn down in 2004. Courtesy of
Fairfax County Library, Photographic Archive

This book is the result of my attempt to answer that question. In the course
of my investigations I found out who many of the tenants of those buildings
were and what they did. I found out the purpose of the radio tower, discover-
ing not only that it had little to do with the numbers transmissions, but also
that it is not central to what does go on in Tysons Corner. Still, my initial
hunch that whatever was going on had something to do with classified military research was correct, even if that hunch was naive. What I found turned
out to be a richer, more complex, and far more interesting story.

I found that this compact suburban region represents a distillation of a
number of fascinating topics. One concerns federal policies supporting scientific research in the service of national defense during and after the Second

World War. Another, revealed by the names of the tenants on top of those buildings, reveals how the government's role in that research shifted from one dominated by federal laboratories to one dominated by contracts with private companies. The pattern of office park and retail development tells us about postwar policies of highway construction, the movement from central cities to suburbs, local politics of land use and government, and a shift from street-cars and railroads to automobiles and airplanes. And all of that took place in a region known to Virginians as hallowed land, one which saw conflict during another war—the Civil War. The tenants who occupy Tysons Corner build-ings do not think about the Civil War very much; perhaps only to note the occasional reenactments near the local battlefields of Manassas or Antietam. But issues of armed conflict are never far from their work.

A few miles east of Tysons Corner on Arlington Ridge Road, the Common-wealth of Virginia placed a historical marker commemorating the location of the estate of a local resident named James Roach. According to the plaque, his property was "ruined and vandalized during the construction of Fort Runyon and Fort Albany in 1861." These were forts the Union army built to defend the capital from a Confederate invasion. The plaque is within sight of the south facade of the Pentagon, which was struck by an airplane hijacked by terrorists on September 11, 2001. Tysons Corner owes its existence to the nearby loca-tion of the Pentagon, and Tysons Corner will continue to support Pentagon activities as long as there is a need for national defense.

What follows is what I have uncovered. The following chapters give equal weight to both themes: on the one hand, national defense, and highways, land use, and suburbanization on the other. Both are critical and one cannot properly study the region otherwise.

ACKNOWLEDGMENTS

Among the many people who have helped me with this study, I wish to give special thanks to the following: Donald Baucom, Andrew Boncek, Christopher Bright, Roberta Buchannan, Peter Chapin, Lisa Davidson, Laura DeNardis, Gregory Dreicer, Ed Durbin, Steve Fuller, Joel Garreau, Martina Hessler, Peter Katz, Thomas Lassman, Jennifer Light, Joan Mathys, and Heike Meyer. Portions of this study were presented to the Division of Space History at the National Air and Space Museum, the Society for the History of Technology, the Midwest Junto of the History of Science Society, the International Congress of the History of Technology, the Society for History in the Federal Government, and to the Washington chapter of the Society of Architectural Historians, to whom I am grateful for their comments. I was also assisted by the staff at the Library of Congress, the National Archives, and especially the Virginia Room of the Fairfax County Library. I would also like to thank my family for their patience and support during the time I worked on this book. The opinions expressed herein are the author's.

Years ago I heard a story, possibly a variant of Aesop's Fable "The Fly and the Draught-Mule," in which a fly sitting on the axle of a chariot exclaims to the mule who is pulling it down the road, "What a mighty cloud of dust we are kicking up!" The mule knows better. This study examines a chariot hurtling through northern Virginia, one whose economic vitality and dynamism are the envy of the world. The entire Washington, D.C., region has experienced this growth, but it has been concentrated across the Potomac in Fairfax County, Virginia, spilling over into neighboring Arlington and Loudoun counties and the city of Alexandria.

Who were this region's drivers as it became one of the biggest centers of employment and shopping on the East Coast, and the engine of the regional economy? In the following pages we will meet a number of individuals—real estate developers, scientists, engineers, business executives, federal and local government officials, as well as several Presidents—who all kicked up their share of dust. Yet it seems that the real driver has been, in the words of one critic, "Uncle Sam.[1]" If that is the case, perhaps northern Virginia would have grown anyway simply because of the presence of the world's most powerful nation's (post-1945) seat of government, a government that has grown in power and complexity without pause for the past 60 years.

That explanation is only half right. After 1945, urban areas throughout the country grew, in some cases much more rapidly than in the Northeast. As was the case elsewhere, much of the growth happened in suburbs. Likewise, these suburbs displaced farms that had been supplying food to city dwellers. Yet what developed in northern Virginia was different. More precisely, something unique happened in a portion of Fairfax County, especially near a crossroads that, into the mid-1950s, was still characterized by dairy farms and

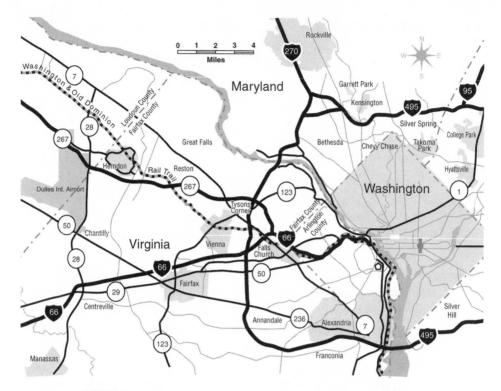

FIGURE I.I

Washington, D.C. and the northern Virginia suburbs. The county of Arlington was the result of the 1846 retrocession of part of the District of Columbia back to Virginia. Tysons Corner lies just to the west, within Fairfax County. Fairfax, Falls Church, and Alexandria are independent cities with their own governments. Credit: ASAP Graphics, Inc.

gravel pits. That crossroads was called Tysons Corner, and its history is the focus of this study.

Tysons Corner is a compact area of intense commercial development, about four square miles in area, located ten miles west of the seat of the federal government in Washington. Fairfax County, in which it is located, is the first major Virginia county south of the Potomac (the smaller county of Arlington lies in between; its unique history will be discussed later). Tysons Corner is unincorporated, unlike nearby towns that have their own government, including Falls Church, Vienna, Fairfax City, and Alexandria. The activities characterizing Tysons Corner—namely, military contracting and systems engineering—take place throughout the region, including in the Maryland suburbs and in the District of Columbia. Nevertheless, what emerged in Tysons Corner is different enough in character to deserve a closer look.

In the 1980s, the main activities of Tysons Corner outgrew the constraints of the four-square-mile area and began to spread westward, out to and beyond Dulles Airport. That development created a linear corridor, about forty miles long, that now begins in Arlington and ends just beyond the airport at the former farming town of Ashburn. A spur of this corridor extends south to the old crossroads of Chantilly, once recognized only by Civil War buffs. Businesses in the western part of this corridor list "Dulles" as their mailing address, and the press often refers to it as "the Dulles Corridor." I use the term "Internet Alley" to distinguish it from Tysons Corner (this is discussed in more detail in chapter 7). When dealing with general issues of technology and suburban growth, one need not distinguish among these regions; hence the more general term "northern Virginia," which refers to most of the Commonwealth situated north of the Rappahannock River. At other times the distinction is important, and that will be made clear in the context of what is being described. I have already hinted that one characteristic of this region is its concentration of government contractors whose work is of a scientific and technical nature, many of them for the Department of Defense and other federal agencies concerned with national security. Following chapters will examine the nature of this work in detail; they will also examine the question of what it is that these contractors do that is truly innovative, in both a technological and a managerial sense.

As I was carrying out this study, that work dramatically changed. The Cold War, whose funding drove Tysons Corner's initial growth, ended, but the transformation of the landscape continued at an even greater pace. The development west of Tysons Corner out to Dulles repeated the established pattern of sleek office buildings housing high-tech tenants, but with a twist. Many of the tenants there were not defense-oriented, but part of the new Internet and telecommunications world. Partly by accident and partly by design, the Dulles Corridor became a world center for the management and operation of the Internet. Security was tight inside these buildings, too, but to protect economic assets, not military secrets.

Real estate developers did not worry about that distinction; they coined the term "Techtopia" to encompass the entire region where all high-technology activity, whether military or civilian, was going on.[2] They made no distinction between Tysons Corner, the Dulles Corridor, or the commercial districts in

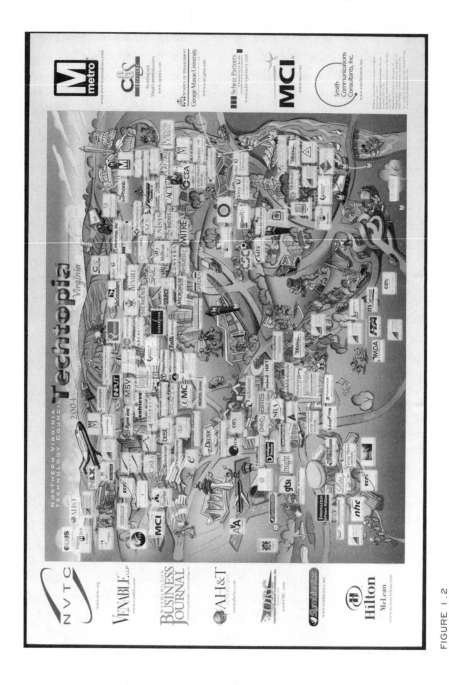

FIGURE 1.2

Techtopia, 2004. The Northern Virginia Technology Council has been publishing these maps annually for a number of years.
Source: Northern Virginia Technology Council.

Alexandria or other nearby towns. Tenants in the Dulles Corridor buildings, however, saw themselves differently: they were funded by venture capital, not by Pentagon contracts. They did not look to hire retired military officers with contacts inside the Pentagon, or care that their offices were physically located close to the Pentagon building in Arlington. These companies kept trade secrets but otherwise were not shy about telling the world what they were doing. The run-up of Internet stocks on the NASDAQ through the 1990s seemed to confirm their assertion that the "old economy," symbolized by many Tysons Corner firms, was obsolete. Were these young entrepreneurs right? They seemed certain of that, although some of those who built the Tysons economy disagreed.

As this transformation to a venture capital-driven economy was under way, two critical events happened. In the spring of 2000, the Internet bubble burst. One local company, MicroStrategy, became a symbol of the bubble as its stock plummeted from $227 to $87 a share on a single day in March.[3] Other Internet and telecom start-ups followed MicroStrategy into trouble. Two telecom giants located along the Dulles Corridor, MCI-WorldCom and AOL-Time Warner, found themselves in financial difficulties as well, and those difficulties became the unwelcome subject of daily business reports.[4] The Internet-based "new economy" was not so robust after all.

Then, on the morning of September 11, 2001, terrorists hijacked four airplanes and used them to attack symbols of American power. One of the four planes crashed into the Pentagon, killing 125 people in the building and 59 on the plane.[5] The distinction between companies relying on military contracts and companies supplying commercial Internet and telecommunications services was no longer important. It was not the economy of northern Virginia but the security of the United States that was at stake. As the war against terrorism relied on timely information and intelligence, people recognized that the computer and telecommunications firms' work was vital to national defense as well. The events of 2000 and 2001 did not alter the character of the region other than to accelerate the area's already rapid growth. However, the question of how much the federal government was spending as compared to the venture capitalists is no longer an issue, as federal spending increased dramatically after September 11.

GEOGRAPHY

Does it matter that this work happens to be located in a particular place in northern Virginia? What about the similar firms located in Maryland and the District of Columbia? Local residents often wax at length about the differences between Maryland and Virginia. Residents of one state declare that under no circumstances would they consider living in the other.[6] (For a long time both neglected the District of Columbia, which lies between them.) Residents of the Maryland suburbs, for example, think of northern Virginia as a foreign land, because, among other things, it is in the South. This belief, while common, cannot go unchallenged. That same belief asserts that Maryland is a solidly Democratic state, while Virginia, along with rest of the once-'Solid South,' broke away from the Democratic Party thirty years ago and is now Republican. The fact that recently Maryland has had a Republican governor while Virginia's was a Democrat did not change this view, but is a sign that those differences are exaggerated. Both Maryland and Virginia are south of the Mason-Dixon line, both were slave states, and until recently, both grew tobacco as one of their major crops.[7]

Virginia seceded from the Union in April 1861, while Maryland did not. Some of the differences between the two states today are a product of decisions made by the Union and Confederate governments that year. The Potomac River was the border between the United and the Confederate States. However, the Union army occupied the town of Alexandria and adjacent land immediately across the Potomac for most of the Civil War.[8] The portion of Virginia between the Potomac and the Rappahannock was the setting of numerous skirmishes, raids, and battles, but the Confederate government in Richmond made the Rappahannock, not the Potomac, its line of defense against the Union armies. Virginia north of the Rappahannock, the location of present-day Techtopia, was in effect ceded from the Confederacy (although not completely, as shown by the two battles fought at Manassas in 1861 and 1862). Obviously much has changed since 1861, but the current focus of northern Virginia's economy on things military began with that Union occupation of northern Virginia, however unwelcome it was at the time.

A few Civil War skirmishes deserve brief mention. A Confederate raid on a railroad train leaving Vienna on June 17, 1861 marked the beginning of armed conflict in northern Virginia.[9] A few days later, Thaddeus S. C. Lowe tested the concept of aerial reconnaissance for the Union army by ascending in a balloon launched from Falls Church to see if he could spot the locations of Confederate troops. It was the first military use of aerial reconnaissance in the United States. Lowe's balloons flew over land where, today, one finds the headquarters of the Central Intelligence Agency, the National Reconnaissance Office, and the National Geospatial-Intelligence Agency.[10] Lowe's experiments were followed the next month by the First Battle of Manassas (Bull Run), where Confederate General Thomas J. Jackson stood "like a stone wall" against the Union troops. After that defeat, the Union Army retreated while the Confederates built defenses at Centreville to prevent a further incursion.[11] Loyalties among the residents of Fairfax County were divided, although in May 1862 the county government was reconstituted and professed its loyalty to the Union cause. The occupation was chaotic, especially after another defeat at Manassas in August 1862 left many residents of the county outside Union lines. Following that battle, on September 1, 1862, Stonewall Jackson's troops attacked the Union army as it was retreating back to Washington. The armies fought to a stalemate in a brief but violent battle at Ox Hill, west of the Fairfax Court House and about eighteen miles west of the White House.[12] Ox Hill, sometimes called the Battle of Chantilly, was the only major Civil War battle in Fairfax County. Rebel armies would never advance as close to Washington from the south again.[13] Mindful of the defenses around Washington, Robert E. Lee moved his troops to the northwest, crossing the Potomac thirty five miles upstream and engaging Union armies at Sharpsburg, Maryland, on the banks of Antietam Creek.[14] Back in Fairfax and Loudoun counties, periodic raids by Confederate Colonel John Singleton Mosby, the "Grey Ghost," made life miserable for those with Union sympathies. Today's U.S. Route 50, flanked by townhouses, defense contractors, and shopping malls, is named the John Mosby Highway in Loudoun and the Lee-Jackson Memorial Highway in Fairfax County to commemorate those days. In the 1990s, the Ox Hill battlefield was covered by townhouses and the Fairfax Towne Center shopping mall.[15]

After Lee's surrender at Appomattox in 1865, the Civil War became the "Lost Cause," and for many Virginians history itself came to an end, perhaps to reawaken at some unspecified time in the future.[16] A professor at Virginia Tech, seeking to document the history of digital computer usage at a navy yard at Dahlgren, was told by the Virginia State Antiquities Commission that "anything postdating 1865 is not considered historical."[17] During the war,

residents of the Loudoun County village of Waterford had been unsupport-
ive of secession; after 1865 and well into the twentieth century they found
themselves snubbed by the Virginia legislature when it came to appropriations for roads and public works. That had the unintended but fortuitous
consequence of stopping material progress in the village, thus rendering it
now one of the most desirable places to live among wealthy refugees escaping
the hectic pace of Washington life.

Included in the area that remained under Union control was a crossroads
called Peach Grove, just east of the site of the June 1861 raid. Later it became
known as Tysons Crossroads, then Tysons Corner, after William Tyson, a
Marylander who purchased property there in 1852.[18] Some of the earliest
written records of the Tysons Corner economy come from farmers who
petitioned the government for money to compensate for livestock, grain,
and lumber that Union armies took before the battles at Manassas.[19] Peach
Grove became an important source of firewood and lumber, which left the
area stripped of trees. At an elevation of 500 feet above sea level, it was one
of the highest points in the county and to this day affords spectacular views
of the Blue Ridge to the west and the Washington Monument to the east.
The Union army erected a 60-foot signal tower there, using timber taken
from a nearby farm. Using a system of signal flags, Union troops moving to
the west could communicate with troops stationed in the forts surrounding
and defending Washington. Ninety years later, at the beginning of the Cold
War, a U.S. Army radio tower was erected on the same spot. It used radio
technology instead of flags, but for the same purpose: to communicate
between Washington and outlying military posts. For a time it fell into disuse
but lately it has been refurbished and is being used again.

Time flows in a steady, linear march, but to residents of Tysons Corner the
tumult of 1861–1865 was followed by a long succession of days in which
life went on at a slow pace. This time dilation persisted until October 1957,
when the faint beep of the Soviet Sputnik suddenly made everyone's
clocks run faster. A famous photograph of Tysons Corner taken in 1956
shows a pickup truck parked outside the Crossroads Market.[20] The truck's
door is ajar. One can imagine the driver—perhaps someone who worked on
a local dairy farm—taking his time as he conversed with the store's owner or
other customers. The farms had become mechanized, but otherwise the conversations might have been familiar to residents of the area a century earlier.
But that era was rapidly passing. A few years after Sputnik, the crossroads was

FIGURE 1.3

Tysons Corner, 1956. The Crossroads Market stood on the southwest corner of the intersection of Routes 7 and 123 until the intersection was reconstructed in 1963. Thirty years later, the headquarters of America Online (AOL) and the Internet switch MAE-East were located a few hundred yards to the south of this store. Courtesy of Fairfax County Library, Photographic Archive

replaced by an overpass and the stores at the old corner were torn down. Today, those who park in the lots nearby are careful to lock their car doors.

After 1865, the main activity of the northern Virginia economy was once again farming. Cotton and tobacco had never been the principal crops; this was an area for raising dairy and beef cattle, for growing grain and fruit, and for lumber. Fairfax County's dairy farms grew to meet the needs of nearby Washington, D.C. By the early twentieth century, no Virginia county produced more dairy products than Fairfax. The Alexandria, Loudoun and Hampshire Railroad, the target of the 1861 Vienna raid, was rebuilt to serve the growing economy and was renamed the Washington and Old Dominion Railroad. It was often in financial trouble but remained important to the region's farming economy. The alignment of the W&OD railroad is now paralleled by four limited-access highways that define the Dulles Corridor: Interstate 66, the Dulles Airport Access Road, the Dulles Toll Road, and the

Dulles Greenway.[21] The railroad never pierced the Blue Ridge or reached the coal fields of West Virginia, but it did survive long enough to deliver loads of construction material for Dulles Airport, completed in 1962. The line was abandoned in 1968, its right of way converted to a linear park that is now among the most popular rails-to-trails paths in the country.[22]

One other change to Fairfax County after 1865 needs to be mentioned. That was the end of the custom of locating political power in the county court. Many of the court's roles were assumed by a board of supervisors, consisting of elected officials from newly divided sections of the county.[23] This form of government would play an important role in the commercial and residential development of Tysons Corner in the modern era.

THE BEGINNINGS OF SUBURBANIZATION

Northern Virginia farmers did well after the Civil War. Despite the image of predatory northern carpetbaggers, the region benefited from an influx of northerners, some of them Union veterans who remembered the lands they had crossed. They brought with them much-needed capital, which they used to buy land, make improvements, and establish prosperous farms and accompanying mills. The proximity to Washington was a crucial factor in establishing successful farms and businesses. Because Fairfax County was still far removed from Washington, the roads were of poor quality, and the river crossings were few, suburbanization could not yet take hold. The W&OD Railroad and the Leesburg Pike (present-day Route 7) were both good transportation routes, but they terminated in Alexandria and did not lead directly to Washington. An old road from Fairfax led through Vienna east to the river, crossing the Potomac at Chain Bridge. It crossed the Leesburg Pike at Tysons Corner and is now known as Route 123. But it went over difficult terrain, and the river crossing at Chain Bridge was still a fair distance from downtown Washington.[24] A group of speculators purchased land along the W&OD near Vienna around 1887 and began selling residential lots, but their attempt to establish the community of Dunn-Loring (named after two of the group) was premature.[25]

Commuting became more practical at the turn of the twentieth century, when a number of electric streetcar lines were built. These were specifically intended to create suburbs from farmland and to serve commuters. Rather than head for Alexandria, these lines aimed at Washington, D.C. and crossed the Potomac farther upstream. Lines from Alexandria and Mount Vernon crossed the river and reached the District over a bridge near the present-day Jefferson Memorial. Lines from Fairfax, Vienna, and the western parts of Arlington converged at Rosslyn, just across the river from Georgetown. Rosslyn became a busy terminus; from there, commuters would cross the river to a de facto Union Station (still standing) at 36th and M Streets in Georgetown. There they could transfer to lines into downtown Washington. At the initiative of Senator Steven B. Elkins of West Virginia, a line was laid out from Rosslyn to the Great Falls of the Potomac, the region's most spectacular natural feature. Excursion trains to the falls began running in 1906, while closer in, the line established the suburb of McLean.[26] Portions of the W&OD were electrified in 1912, and a branch that diverged to Rosslyn began to serve commuters as well as freight customers. The village of Clarendon and the towns of Falls Church and Fairfax all had electric rail connections to Washington by 1904.

The lines did not last long but they had a profound effect. They began the transformation of Northern Virginia from a farming region to a bedroom suburb. After a method for state financing of road building was established, it became practical to commute by automobile as well, and the combination of automobiles and publicly-financed roads doomed the streetcar. The line from Mount Vernon was abandoned in 1932, the Great Falls line in 1934, and the Fairfax in 1939.[27] The Washington & Old Dominion dropped electric-powered passenger service in 1941 and reverted back to a freight line, its trains pulled by diesel locomotives.[28] But while automobiles drove the electric streetcar lines into bankruptcy and abandonment, the streetcars set the pattern of suburbanization, with its subdivisions of farms into residential lots. In many instances the roads took over the railroad's right of way; for example, Old Dominion Drive through McLean.[29] Nearly all other traces of the streetcar lines have vanished except for a few street names that reflect their passage, such as Electric Avenue in Vienna, or Railroad Avenue in downtown Fairfax.

In 1846 Virginia petitioned the federal government to take back the part of the District of Columbia that lay on the Virginia side of the Potomac. Retrocession of what was then farms and a few small villages took effect the following year, and the area was incorporated into Alexandria County. In 1920, with suburbanization underway, the thirty three-square-mile parcel was incorporated and named Arlington County, after the name of Robert E. Lee's home before the Civil War. The 1846 decision was based on the perception that too much land had been taken for the new capital; the 1920 incorporation of Arlington County served as notice that the original 1846 decision, however flawed, would not be reversed. By 1920, thanks to new river crossings and streetcar lines, Arlington County was a fast-growing suburb and home to many federal workers. Old farming or streetcar commuter villages like Clarendon lost their individual identities and merged into a continuous spread of housing developments, served by roads leading to bridges across the Potomac. Arlington thus acquired the unique character of being physically a part of Virginia yet perceived—correctly—as being culturally, politically, and economically more a part of Washington.[30] It retains its uniqueness to this day: neither a city (like its neighbor Alexandria) nor a county in the traditional sense. Since about 1980, dense urban development has begun to crowd out some of the inner suburbs, although large areas of Arlington County retain a bedroom community flavor. It retains a system of county government.

THE SECOND WORLD WAR AND REGIONS OF HIGH TECHNOLOGY

The preceding discussion of the Civil War, Reconstruction, and the region's economy brings us back to the original question of what qualities, if any, distinguish northern Virginia from similar areas in suburban Maryland or the District of Columbia. The pundits who write for the op-ed page of the *Washington Post* and who populate the Sunday morning talk shows insist that there is a big difference from their vantage point. Northern Virginia has many qualities that are the true reasons for the mix of military and civilian technology firms, residences, and retail shopping now located there. The cultural differences between North and South, though present, are not among them, but they are worth some examination.

To Virginians who study the early history of their state, the phrase "the War" means one thing: the conflict with the North between 1861 and 1865. In the course of researching this study, I interviewed executives, real estate developers, scientists, and engineers whose work gives the region its identity today. These men and women also spoke of the enormous effect of "the War," except they meant the Second World War, fought between 1939 and 1945, not the Civil War. World War II affected all regions of the United States, but it affected the South perhaps more than others. And within the South, it affected northern Virginia most of all. World War II changed the relationship among the military, armaments manufacturers, and contractors. It changed the role of the scientist in society and the role of universities. It helped create what President Eisenhower called, in his 1961 farewell address, the "military-industrial complex."[31] There were differences of orders of magnitude between the Union signal tower at Peach Grove and the microwave tower that replaced it in the 1950s, or between Thaddeus Lowe's balloons and the reconnaissance satellites of the 1960s.

During the Civil War and Reconstruction, Virginians saw the federal Army as pillagers, who helped themselves to livestock, grain, and wood as they needed it, burning the rest. After World War II, northern Virginians saw the U.S. military establishment as a source of both pride and employment, as it established numerous army and naval bases in the state. In 1942 the War Department moved its headquarters from Washington to Arlington, to a five-sided building called simply the Pentagon. That move was not seen as an occupation this time.

Students of southern history often say that "the South" really begins at the Rappahannock, the next major river south of the Potomac, thirty miles from the Maryland-Virginia border. That was the de facto border during the Civil War, and it has persisted as a boundary into the twenty first century. As late as the 1980s, drivers entering Virginia from the north on Interstate 95 first saw a "Welcome to Virginia" sign at the Rappahannock, not the Potomac River bridge.[32] Students of geography have other tests for whether a place is in the South or not: they go into a restaurant and see if grits are on the menu. My own test is similar: if diners sitting in the next booth mention "the Yankees," are they discussing Lincoln's armies or the baseball team from the Bronx? By these criteria, Tysons Corner and the Dulles Corridor are not in the South.[33]

The other types of differences are those that primarily led to the concentration of certain types of firms in the Dulles Corridor, while other types developed in the Maryland suburbs. One must avoid the historian's fallacy of saying that because such a concentration *did* develop in Virginia, it *had* to have developed there. In hindsight it is easy to pick out the factors that led to the asymmetry, but anticipating those factors is hard. That does not prevent city planners and politicians from around the world from visiting northern Virginia, hoping to find the key to its concentration of wealth and high-paying jobs and take it back to their countries or regions.[34] These visitors typically have lunch at the Tower Club in Tysons Corner, visit office buildings in Reston and Herndon where major Internet work is done, visit the offices of America Online in Dulles, and schedule interviews with local entrepreneurs, real estate developers, and executives, who are all happy to tell them the "secret" to the region's success.

Before they come to Virginia, these visitors often go first to California, where a concentration of semiconductor, computer, and software firms arose in a compact area on the peninsula between Palo Alto and San Jose. Around 1970, a local journalist gave that area the name "Silicon Valley," a term that has come to symbolize prosperity based on information technology. Like Tysons Corner and the Dulles Corridor, Silicon Valley is not an incorporated town. There is no post office with that name, nor is there a mayor or city council that governs it. Unlike northern Virginia, it has been the subject of numerous academic studies, yielding shelves full of books and reports on its origins, growth, vitality, and future. Returning to their home regions, people set out to create a local version, which the local press dubs "Silicon X," where X stands for Glen, Gulch, Plains, Forest, Beach, and so on.[35] From these studies one can distill a consensus on what factors are critical for success:

1. The presence of a nearby highly ranked research university with strengths in computer science, electrical engineering, and the physical sciences
2. An entrepreneurial climate including the presence of venture capital and forward-thinking banks
3. An infrastructure of cooperative local zoning boards, real estate developers, lawyers, marketing people, and the like who can help a company get established quickly without red tape
4. Suitable places for engineers and their families to live, with good schools, roads, and parks
5. The presence of a military or other federal research facility to pump in government money

This last criterion apparently contradicts the second, which implies that private sources of capital are better engines of growth than government money, since the former is more flexible, more attuned to market needs, and less influenced by unseemly political pressures. As we study those factors in relation to Tysons Corner and the Dulles Corridor, we need to keep that difference in mind. Robert Noyce, the coinventor of the silicon chip and one who helped give Silicon Valley its reputation, remarked that military-supported projects were "a waste of the asset. The direction of the research was being determined by people less competent in seeing where it ought to go, and a lot of time of the researchers themselves was being spent communicating with military people through progress reports or visits or whatever."[36] In the culture of Silicon Valley, it is always the engineer, the programmer, even the computer hacker, who ranks at the top, even if he or she may not be the CEO of the company or necessarily have gotten rich from his or her efforts.[37]

It is worth stating that northern Virginia easily qualifies as a technology center, with a concentration of high-paying jobs. Northern Virginia's economy is dynamic and fast-changing, and it is impossible to cite exact figures. But a survey of technology employment conducted in 2000, the year the Internet bubble burst, listed the Washington region as having 426,000 tech jobs, third behind Silicon Valley (629,000) and Boston (526,000). It had nearly twice as many as the next place on the list, Seattle (234,000).[38] That survey looked at the entire Washington area, but other surveys have shown that about two-thirds of that work is done in the Virginia suburbs— overwhelmingly in the Tysons-Dulles axis, with the rest in Maryland and only a small amount in the District.[39] Thus northern Virginia would remain third-ranked in the 2000 survey. Other surveys have measured variations of the local economy taken before, during, and after the Internet run-up; they all present similar conclusions. Within the Commonwealth of Virginia, the region near Washington is without question the economic engine. As early as 1989, then-Governor Gerald L. Baliles noted that Virginia income tax receipts from Fairfax County alone exceeded revenues from the cities of Richmond, Norfolk, Virginia Beach, Roanoke, Winchester, and Bristol, plus the counties of Albermarle, Botecourt, Chesterfield, Halifax, Rockbridge, Tazewell, and Westmoreland *combined*.[40] Other informal surveys of the Washington region state that from 40 percent to 60 percent of Virginia's tax revenues come from an area less than fifteen miles from the Washington Monument. If this region is less dynamic and prosperous than Silicon Valley, it cannot be far behind.

Of the factors listed earlier for Silicon Valley, some are present in this region, but others are conspicuously absent. The Washington, D.C. area is blessed with a number of high-quality universities and colleges, but the one that is best known for computer science and electrical engineering is the University of Maryland, on the other side of the Beltway in College Park. Virginia's technical university, Virginia Tech, is highly-ranked but is in

Blacksburg, 250 miles away. George Mason University in Fairfax has built up comparable strengths in these subjects, but that university was created in response to demand from local firms—it is an effect, not a cause, of the booming economy. Nearby universities in the District of Columbia, including Georgetown, George Washington, and Howard, are top-ranked but are better known for their strengths in medicine, law, and politics than for computer science, physics, or electrical engineering.

Among the other factors, northern Virginia fares well: it has a pleasant climate, access to recreation, and good schools. The local governments and business community have been supportive, although not without frequent rear-guard fights from local residents' associations. Like Silicon Valley, it has a good infrastructure of roads, airports, water, sewer, and electric utilities. (Also like Silicon Valley, these highways can no longer handle the traffic.) Northern Virginia's telecommunications infrastructure may be the best in the world, although that, like George Mason University, is more an effect than a cause of economic development in the region.

Federal facilities in northern Virginia, however, are nothing like what one finds in Silicon Valley, and indeed nothing like any other region of the country. Assessing their role requires a careful and detailed analysis, and that will be a major focus of subsequent chapters. The difference is one both of degree and of kind. Military and federal research laboratories of all sizes and types are scattered throughout the District and around the Washington area. (In the immediate suburbs, more of these laboratories are located in Maryland than in Virginia.) The region is also home to institutions serving the management of research, engineering, and military activities. These are not unique to this region, but they are not found in such a concentration anywhere else. And whereas one can find military bases and government laboratories in Virginia, Maryland, and the District, these entities dealing with management are heavily concentrated in Virginia, where their presence gives Tysons Corner its unique character.

The critical role of the Defense Department in northern Virginia affects the other criterion for high technology economies: the requirement that a region have risk-taking bankers and venture capitalists, with their preference for lean and flexible organizations that keep stifling bureaucracy at bay. Robert Noyce's comments suggest that Silicon Valley would not exist were it not for such risk-takers. How does that translate to northern Virginia? After all, the Pentagon may look like a fortification with its five sides, but it is really just a large office building, housing one of the largest bureaucracies in the world. The military's weapons are built and tested elsewhere; the Pentagon's principal product is paper (or its modern equivalent, PowerPoint slides). Can one base a dynamic, technology-driven local economy on contracts from government bureaucrats? Subsequent chapters will show that the answer is yes. The government's presence, however, has made the nature of the Virginia economy different from that of Silicon Valley, Boston's Route 128, the Dallas-Fort Worth Metroplex, or other similar regions. Venture capital can be raised in northern Virginia, but activities funded by it may never grow out of the shadow of those receiving government monies. This implies that the Dulles Corridor will never achieve the kind of dynamism found in Silicon Valley. Perhaps the nation needs only one such place anyway. The role of government funding, especially Pentagon funding, versus venture capital and Wall Street funding will also come into play as we examine the difference between what drove the growth of Tysons Corner and what grew up along the Dulles Toll Road.

Beginning in the next chapter, we will leave our discussion of the Civil War and Reconstruction, which many Virginians feel is the true subject of Commonwealth history. But it will be worth remembering that the post-World War II development of government contracting, consulting, systems integration, and telecommunications all happened on specific pieces of land. But how does one connect the present to the past in a meaningful way? The townhouses and office towers one finds there today look the same as structures found almost anywhere else in the eastern United States. The shopping malls are filled with chain stores and restaurants using the same logos and fixtures, selling the same clothes, and offering the same menus as anywhere else. Does it make any difference that Stonewall Jackson led his men over what is now the site of the Fairfax Towne Center?[41] It is a nice conceit to compare Thaddeus Lowe's balloons at Falls Church to the National Reconnaissance Office's activities in Chantilly, or the "Quaker guns" (logs painted to look like cannons) deployed at Centreville to the CIA-manufactured deceptions produced in its labs at CIA Headquarters in Langley.[42] But is there anything of more substance to that comparison? We shall return to that question in the following narrative.

The person who unleashed the forces that created northern Virginia was a flinty Yankee, a descendant of Cape Cod sea captains, a conservative who distrusted Big Government (including Roosevelt's New Deal), and is today remembered less for what he did in Washington than in what he did in Cambridge, Massachusetts, where he had been a student, faculty member, and administrator at Tufts and MIT. Before the Second World War, Vannevar Bush had a reputation as a prolific inventor. Among his inventions was the differential analyzer, an analog computer that was a precursor of the computers that would follow at MIT and elsewhere. He also conducted research on radio. To commercialize his innovations in this field, he helped found Raytheon, now a major aerospace and military electronics supplier with over 5,000 employees in the Washington region, and with facilities in Arlington, Vienna, Falls Church, and the Dulles Corridor.[1] Another avenue of Bush's prewar research concerned information retrieval devices, using the then-novel technique of microfilm. For that work, and for an essay on that topic he published after the war, people have called him the spiritual father of the World Wide Web.[2]

Between 1939 and 1955, however, Bush was not at MIT but in an office in Washington. There he served as the head of the Carnegie Institution, a respected patron of scientific research founded by Andrew Carnegie in 1902. Although reluctant to leave his beloved New England, Bush wanted to be more involved in the politics of government support for research. He saw no way for the United States to avoid getting involved in the conflicts breaking out in Europe and the Pacific, and he saw that the United States military was not ready. He already had the reputation and connections to change that, though it did require moving to Washington, D.C.[3]

Bush took charge of the Carnegie Institution in January 1939, moving into its offices on 16th and P Streets, north of the White House. By then he was already the chair of a division of the National Research Council, and he had also been appointed to the National Advisory Committee for Aeronautics (the predecessor of NASA). In October 1939 he was appointed chairman of the NACA's Main Committee, a post he held only briefly as the United States entered World War II.[4] With conflict in Europe breaking out in September 1939, Bush began to champion the establishment of an organization "to co-ordinate, supervise, and conduct scientific research on mechanisms and devices of warfare."[5] He decided that the best way to implement his ideas was to meet personally with the President and get approval from the very top of the government. Through Frederic Delano, a relative of Roosevelt and a trustee of the Carnegie Institution, Bush was able to arrange a meeting in June 1940. The previous month, the Nazis overran the Low Countries. The President's advisers needed little convincing. With Roosevelt's approval, the National Defense Research Committee (NDRC) was established with Bush as chairman and its offices at the Carnegie Institution.[6]

The story of how Vannevar Bush, at one time a vocal critic of the New Deal, became one of Washington's most powerful figures based on FDR's trust in his judgment has been told many times. For this history of the Washington region, we note one facet of the relationship between Bush and Roosevelt: how their personal relationship led to a new kind of institutional relationship, between the federal government on one hand, and academic and industrial research organizations on the other. What was novel about the NDRC (and its successor, the Office of Scientific Research and Development, or OSRD) was that it would contract for research to be conducted at the scientist's or engineer's home institution, as well as supporting research at a federal or military laboratory. Scientists would retain the autonomy they had before the war while their energies would be focused on winning the war. They would not have to give up their faculty positions, but the nature of their work would be set by the NDRC contracts as directly as if they were all suddenly given commissions and ordered to work at a military lab.[7]

In practice, the contracts went mainly to universities, and the best of the universities at that: Harvard, Cal Tech, MIT, Columbia, the University of California. Although the plan was to avoid establishing new federal laboratories, some were established, although not as free-standing entities but instead usually tied to a research university. The most famous was the Radiation Laboratory, established on the MIT campus for work on radar. Locally, George Washington University was a beneficiary, as was Johns Hopkins University of Baltimore. Contracts also went to private, nonprofit research institutions including the Carnegie in Washington, the Woods Hole Oceanographic Institute, Batelle Memorial Institute, and the Franklin Institute. Contracts went to industrial firms including Western Electric, General Electric, RCA, DuPont, and others.[8] But for-profit research institutes, as we shall discuss later, were rare or simply did not exist at the time. Such for-profit entities would emerge in the late 1950s and early 1960s, and when they did they would find that this relationship, forged by Vannevar Bush in the 1940s, would carry over with little modification. Bush's biographer, G. Pascal Zachary, called him the "Engineer of the American Century" for his unique combination of inventive, academic, administrative, and political skills.[9] By establishing what Zachary called "federalism by contract"—the extension of federal power and influence unaccompanied by an increase in civil service employees or government-operated facilities—Bush became the engineer of Tysons Corner as well.[10]

OPERATIONS RESEARCH, RAND

Of the myriad advances in science and technology that came out of this alliance during World War II, one had the most direct effect on the local economy. It was one that Bush himself was skeptical about. It was not the atomic bomb, nor was it radar, nor the computer, but rather a branch of applied mathematics called operations research, or OR. (In the United Kingdom, where it originated, it took the name operational research.) The term is not easy to define. OR grew out of a realization that while scientists were making dramatic advances in the design, development, and production of new weapons, there needed to also be a corresponding application of science to the deployment and use of those weapons in the field—the "operations" aspect of fighting the second World War. The specific impetus was the British development of radar in the late 1930s to defend against a German air attack across the Channel.[11] The radar sets themselves were remarkable inventions but had to be integrated into the whole system of air defense to be effective. That involved a number of activities: training operators to use the new technology, devising a system of communications among the radar stations and with the headquarters of the Royal Air Force (RAF), and undertaking statistical analyses to determine where placing the radar sets would do the most good. Some of those activities were more politics and management than science, although the work involved advanced mathematics. The term itself was coined sometime in 1938, and after the war began in September 1939, operational research groups were an integral part of the RAF's air defense installations.[12]

In the United States, Bush's template for support of science though NDRC, and later OSRD, contracts specifically excluded research of this nature. That began to change in 1940 as British researchers traveled to the United States, bringing with them the breakthroughs they had made, especially a device called the cavity magnetron, which made microwave radar feasible.[13] In addition to bringing the cavity magnetron, British scientists also brought knowledge of and enthusiasm for the OR techniques they had been developing in tandem with the microwave radar itself. Under Bush's leadership, the OSRD established the previously mentioned Radiation Laboratory on the MIT campus to exploit the British invention. Bush remained reluctant to support OR, seeing that OR was as much developing political and social relations between scientists and military commanders in the field as it was doing mathematical or physics research. Nevertheless, by 1942 the science had taken root in the United States. One factor may have been the shock of the Japanese attack on Pearl Harbor on December 7, 1941: that morning Japanese planes had been detected by a radar installation on the island, but that information was not conveyed to those who could have acted on it in time. The champion of OR in the United States was Professor Philip Morse of MIT, who in the spring of 1942 joined an effort to improve U.S. antisubmarine warfare operations. In Morse's words, "To let nonmilitary persons participate in even minor operational decisions was, of course, heretical to many officers, especially those in the Navy, with their tradition that the commander of the ship was absolute master."[14] But the seriousness of the threat to transatlantic shipping posed by German U-boats was enough to cause a rethinking of this belief. With these footholds OR took hold in the United States.

After the war Morse continued his advocacy. Bush's conservative model of focusing government support for science in the universities remained. The armed services supported Bush's vision, but each branch also established a center where OR could be carried out under direct military sponsorship rather than through contracts. The most famous was Project RAND, established in 1946 at Douglas Aircraft in southern California under an Army Air Forces contract. After the Air Force was established as a separate branch of the military in 1947, Project RAND evolved into the RAND Corporation with its own budget, personnel, and a building in Santa Monica. RAND was not strictly speaking an operations research center, but its mission incorporated and extended the philosophy of OR; namely, to use the most advanced techniques of mathematical analysis as a tool for understanding how new weapons could enhance U.S. security.[15] The popular press made the RAND Corporation a frequent topic of its stories about the Cold War, during which RAND acquired the nickname "think tank"—a name that not all RAND employees liked but one that has stuck.[16]

Discussions among RAND's founders raised issues that would be relevant to the suburban Washington story. One was that, although its contract was with Douglas Aircraft, RAND's goal was not to produce hardware (for example, a ballistic missile) but rather studies outlining the challenges and opportunities for future air force weapons development. RAND would study ballistic missile design (then in its infancy, based on experience with German V-2s), but in a larger context of reconnaissance, global communications, basing of air force facilities in Europe or elsewhere, and, not least, economics.[17] RAND's principal product would be reports, not hardware.

In other respects the RAND Corporation was not the model for what developed in northern Virginia after World War II. Its California location was about as far from Washington as one could get in the United States. That happened because Douglas Aircraft was in Santa Monica, but there may have been other reasons for its location. A site closer to the Pentagon might inhibit a think tank's ability to project into the future, which was what RAND was founded to do. But the armed services also wanted scientists nearby, and they established counterparts to RAND located in the Washington region. During the war, the navy had set up a Naval Operations Research Group, which it sought to preserve after 1945. MIT agreed to sponsor it beginning in November 1945, but it was not located in Cambridge. The renamed Operations Evaluations Group was run from an office in the Pentagon.[18] This group expanded throughout the early Cold War and generated spin-offs of similar groups focused on specific problems. Most of these were of modest size, employing on the order of a dozen scientists, and most were located in the Pentagon or in nearby office buildings in Arlington. These were combined into a single Center for Naval Analyses (CNA) in 1962; the result was a navy counterpart to RAND. The navy wanted the Smithsonian Institution to administer it, but that was vetoed by the Chief Justice of the Supreme Court, who was a member of the Smithsonian Board of Regents. The Franklin Institute got the contract instead.[19] Although the Smithsonian had a long tradition of scientific research and was the incubator of modern agencies like NOAA and NASA, by the 1950s it was better known as a public museum—the "Nation's Attic"—and the Regents saw this contract as pulling it in the wrong direction.

The Army's response had a greater impact on the growth of military contracting in the Washington suburbs, even though its initial contracts went elsewhere. The Army created the Operations Research Organization (ORO) June, 1948. It was administered by Johns Hopkins University, but was substantially independent. Initially its offices were at Fort McNair in southwest Washington, and in 1952 it moved to Chevy Chase, Maryland.[20] Its charter was similar to RAND's in that it sought a university-like environment where researchers were free to look at the larger issues. But the Army's requirements, and its location, meant than in practice the ORO worked much more closely on the Army's immediate needs, especially concerning logistics. In 1951 the Army established a Human Resources Research Office, administered by George Washington University, to study what are now known as "human factors" or the "human interface" between soldiers and new weapons. Two years later it established a Combat Operations Research Group at Fort Monroe, Virginia.[21]

THE MATHEMATICS OF OPERATIONS RESEARCH

Operations research encompassed a number of mathematical techniques—some traditional methods of statistical analysis, others new. The original definition from the British scientists who developed it did not mention mathematics at all: "the evaluation of an equipment or weapon to discover how well it performed on operations, and, secondly, the analysis of operations to see how the equipment or weapon fitted the tactics, or, alternatively, to what extent tactics dictated the form of weapon that was chosen."[22] In practice, as it evolved at RAND and elsewhere, practitioners developed specialized techniques that comprised the core tools of the field. Some of them were relatively new developments: Queuing theory, simulation, the Monte Carlo method, and linear programming, to name a few. To discuss these fully would require an understanding of advanced mathematics and is better left to more specialized textbooks.[23] What follows are brief descriptions that will give a flavor of the pioneering work done at these centers. These descriptions will also help answer a perennial question asked about the Tysons Corner companies; namely, "what do these people do, anyway?"

Queuing theory attacks a problem familiar to anyone who has waited in line at a post office, bank, or supermarket. In some cases all the customers wait in a single line, and each is served by the next available agent. Supermarkets and highway toll booths ask that their customers choose a line, which we do based on its length and our guess as to how many groceries the people in front of us are buying. In theory this method should be less efficient, but if there are enough stations it works fairly well. The many variations based on the number of lines, number of customers, time needed to serve a customer, number of servers, and so on are obvious. This concept has a wide range of applications: air traffic control, telephone switching, dispatching of fire and rescue vehicles, and assembly-line manufacturing, to name a few. The theoretical basis for queuing was studied and worked out early in the twentieth century. During and after World War II it was brought to a very high level of sophistication and applied to a number of military operations. As a relatively mature theory, its use was not restricted to the OR groups mentioned earlier. This theory also forms part of the theoretical basis for the Internet, especially for the Internet's ability to scale up to millions of switching nodes without a serious degradation in performance. It is no coincidence that the early concept and analysis of what is today's Internet came from a researcher at RAND, Paul Baran, who was working on military communications in time of war. Similar work was being done in the UK by Donald Davies of the National Physical Laboratory, and the detailed application of queuing theory to the Internet was carried out at UCLA under the leadership of the mathematician Leonard Kleinrock.[24] Although little of this early theoretical work took place in northern Virginia, we shall see that the Internet and Virginia's economy would have a lot to do with each other later on.

Simulation was an activity well known to the aeronautics community. Beginning in the 1930s, the community recognized the need to train pilots to a certain proficiency level before putting them in a cockpit. And no amount of classroom or textbook training could prepare them for the variety of contingencies of flying, for which they had to be trained to act properly. To that end inventors developed training devices that stayed on the ground, yet recreated this dynamic aspect of flight. It is this modeling of the dynamic element of the physical world—the operation of the model over a span of time—that distinguishes simulation from traditional methods of analysis and study.

Around the time of World War II, mathematicians recognized that simulations could be attempted for other physical systems, especially the economy of a nation.[25] After the war the concept of economic modeling over time took root among economists at RAND, and by about 1950 the advances in electronic digital computing promised finally to make this work practical. In the mid-1950s the IBM Corporation developed the high-level programming language FORTRAN, which allowed those not schooled in the minutiae of computer technology to write sophisticated programs. The RAND mathematician (and Nobel laureate) Harry M. Markowitz wrote an extension of FORTRAN that was optimized for simulations, while IBM developed a similar program called GPSS (General Purpose System Simulator). By 1960 there were other systems available, dovetailing with the availability of reliable commercial computers.[26] The roots of these systems were in economics, but they were of such general purpose that they were found useful in a wide variety of fields. Markowitz soon left RAND and cofounded a company that, decades later, remains one of the most important in northern Virginia: California Analysis Center, Incorporated, or CACI.[27]

One aspect of simulation is the need to understand a complex physical process as it takes one of many possible branches into the future. Traditional mathematics can predict the future path of the planets and events like solar eclipses, but many of the problems that concerned operations researchers cannot be so modeled. The Polish mathematician Stanislaw Ulam, working on the design of the hydrogen bomb at Los Alamos in 1948, came up with a technique of introducing variables with random values, following them through a simulation, and observing the resulting physical system after a simulated interval. By performing what Einstein once called thought-experiments, only this time using a computer, the physicists could learn about a system that they otherwise could not analyze. Ulam's colleague Nicholas Metropolis dubbed it the "Monte Carlo method" after the famous gambling city on the Mediterranean (Las Vegas, though closer to Los Alamos, was hardly known in 1948 as a gambling town). In part because of the catchy name and in part through Ulam's extensive promotion, the technique took hold—again coming along just as powerful electronic computers began to appear to carry out the calculations.[28] It soon became one of the most heavily used techniques among weapons designers, war-gamers, economists, operations researchers, and many others. One interesting side effect of Monte Carlo was the need for numbers that were truly random, something mathematicians thought would be easy to generate but turned out to be very difficult to do with confidence.[29] The RAND Corporation took this on as one of its tasks, publishing in 1955 a famous and classic book, *A Million Random Digits with 100,000 Normal Deviates*—a book consisting of exactly what its title stated: digits that bore no relationship to each other![30]

Finally, it is worth describing the mathematical technique known as linear programming, since it, too, was and remains fundamental to OR. Like the methods described above, it also has a strong connection with RAND in Santa Monica, but we shall see that its history also has an interesting northern Virginian component. Despite the name, linear programming is not a part of computer science, although the technique only became practical when digital computers became available. Briefly, it consists of writing out a series of inequalities that are greater than one variable, while minimizing another. One example is the provisioning of food for an army, in which all of the basic requirements for protein, water, carbohydrates, and so on are exceeded while the total weight of the rations is minimized (no one joins the Army for the food). Such problems take the form of large matrices of numbers representing the coefficients of the terms of the inequalities. The inequalities, if plotted on graph paper, would be bounded by straight lines; hence the method's name. In graphical terms, a solution would describe an area bounded by these lines, and then find a point inside that area where a maximal value exists. For practical reasons a graphical technique cannot be used, as the real-world problems contain too many variables to be plotted on two-dimensional graph paper. During World War II a variation of this problem was solved to coordinate ship travel across the Atlantic—the ships delivered supplies to the European theater of war while minimizing travel in ballast (in other words, with no useful cargo). In the late 1940s the air force asked the Stanford mathematician George Dantzig to solve a problem concerning recruitment of pilots, so that the required number of trained pilots would be available where and when they were needed. Working at Stanford and later RAND, Dantzig came up with a technique that guaranteed a solution.[31] He arranged for a test to be conducted on a nutrition problem, using human clerks equipped with mechanical calculators. Its success in 1947 cleared the way for full-scale applications using the electronic computer just then coming online.

Linear programming proved to be as important to military logistics planners as the Monte Carlo method was to physicists. The key to both, however, was access to an electronic computer, which in the late 1940s was not easy to obtain. The Air Force agreed to purchase the second computer off the production line from the only American company building computers for commercial sale, the Eckert-Mauchly Division of Remington Rand, which was marketing a computer it called the UNIVAC. But the UNIVAC's creators were having troubles bringing this new invention into existence. The Air Force turned to the Bureau of Standards, where an interim computer was being built for experimental use until the UNIVAC could be delivered. This computer, named "SEAC" for "[Bureau of] Standards Eastern Automatic Computer," was completed at the Bureau's labs in Washington in May 1950, and was one of the first stored program computers to operate in the United States. Despite its limitations, the Air Force was encouraged by the machine's ability to run logistics problems. Two years later, the Air Force accepted delivery of a second UNIVAC and installed it in the Pentagon. Like SEAC, it was also a milestone, as it was the first to leave the Eckert-Mauchly factory and be delivered to a customer on site.[32] Both were pressed into service on linear programming problems for the Air Force. Although not recognized at the time as a milestone, when UNIVAC #2 was installed at the Pentagon it introduced electronic computers into northern Virginia.

THE PENTAGON

I have begun to use "the Pentagon" as a synonym for the U.S. armed services, but we have not examined how that phrase came into use. Before returning to the story of military contracting, we need to visit the 1941 decision to locate the War Department (later the Department of Defense) in that building across the Potomac in Virginia.

The Pentagon is a metaphor for American military power, and along with the White House and Capitol, it is one of the most recognized buildings in the world. More than sixty years after its construction, it remains one of the largest office spaces in the world. It sits on land that was retroceded from the District of Columbia to Virginia, with the result that the Defense Department is one of the few cabinet-level federal agencies not located in the land set aside by the Congress for the "seat of the government."[33] In 1942 the Pentagon and its sur-roundings, though in Virginia, were placed under the political jurisdiction of the federal government.[34] Most telephones in the Pentagon use the Virginia 703 area code, but mail addressed to Pentagon offices carries a 20xxx zip code, not the 22xxx zip codes of Virginia. Those who work there are bound by fed-eral, not Virginia, law.[35] The previous chapter discussed how Arlington was unique among Virginia counties and is not really a county by most def-initions. That applies doubly so to the Pentagon.

The reason for locating the War Department in Arlington was simply that there was not enough easily developable land in the portion of the District that had been set aside for it, and with the onset of war there was no time to assemble a suitable parcel elsewhere in the city. Just prior to the German invasion of Poland in 1939, the War Department had moved from a building next to the White House to a temporary structure called the Munitions Building, one of many World War I-vintage "temporaries" crowding the National Mall at the time. In 1939 the armed forces were still modest in number, but that began to change drastically the next year, as the army increased the number of soldiers five fold, and was looking at expanding even more. Besides the Munitions Building, War Department offices were scattered in twenty buildings in the District as well as in Arlington and Alexandria.[36] A new, permanent headquarters was being built at 21st and C Streets, Northwest, but by the time it was completed in June, 1941, Secretary of War Henry L. Stimson realized that it was already too small.[37] The frantic search for space in the midst of rapid mobilization—which would get even worse after Pearl Harbor—was felt through-out Washington. David Brinkley, the well-known journalist, wrote about how Washington was a "languid Southern town" rudely awakened by the oncoming war, for which it was ill-prepared.[38] Apparently Washington is always a "sleepy Southern town," periodically awakened by outsiders mov-ing in as a result of external events: the Civil War, the Progressive Era, World War I, The New Deal, World War II, John F. Kennedy's New Frontier, the Reagan Revolution, and so on.[39] Nonetheless, the sense of near panic in the city as the Nazis overran Europe and threatened Britain was real.

In May 1941 Army Chief of Staff George C. Marshall began to consider a site across the Potomac, in Arlington, for a series of temporary buildings to consolidate War Department offices. The federal government already owned much of the land, some of which had been owned by Robert E. Lee. Congress supported Marshall's consolidation plan, but wanted the buildings in the District. Marshall pressed on, however, noting that the Arlington site was only minutes from the White House. He rejected sites in Maryland as being too remote, and it would have taken too long to assemble a large enough parcel of land in the Foggy Bottom section of Washington where the Department was supposed to be. The task of executing the plan fell to Brigadier General Brehon B. Somervell of the Army Corps of Engineers, the chief of construction for the War Department. Somervell enlisted the support of a Virginia congressman, and in July he made the bold proposal to build on the Arlington site what would be the world's largest office building, capable of accommodating 40,000 workers. (Arlington County's entire population in 1940 was 57,000.)[40] On Thursday, July 17, 1941, he summoned a group, among them Col. Leslie R. Groves who would later head the Manhattan Project, to provide detailed plans for the building by Monday morning. They worked through the weekend and on Monday morning presented a drawing of a broad, five-sided, low building, whose shape was partially determined by the suggested site and the routing of local roads. Five-sided structures were common in military fortifications going back to antiquity, and that gave the building a unique, instantly identifiable shape.

When the plans were made public they were denounced by citizens and government officials alike, including those who felt that Virginia was no place for such an important office. Somervell's political skills were put to the test. Criticism from Gilmore Clarke, chair of the D.C. Commission on Fine Arts, and from Frederic A. Delano got to FDR himself. Roosevelt liked the building's design but agreed to look at the site and address Clarke's objections that it was too close to the Mall monuments and to the sacred grounds of Arlington Cemetery. In August 1941 FDR got into an open car and was driven across the river to the proposed site. Clarke was in the car and made his case against the site. Somervell was in the car, too, and he made it clear that he did not want anyone meddling with his plan. That prompted a reminder from Roosevelt that he was the Commander-in-Chief of the Army and therefore Somervell's superior.[41] Roosevelt picked an alternate site that Somervell had previously rejected—a low, swampy area then occupied by Hoover Airport, which had just been abandoned after the opening of Washington National Airport in June. But otherwise Somervell's plan for a broad, low building—which remains to this day among the world's largest and most famous—remained intact. To Clarke's and others' disappointment, so too did the decision to build it in Virginia, not the District. Congress appropriated the money and construction began in September 1941. Roosevelt followed the design and construction closely. Among the recommendations he made were that no marble be used in its construction, and that "separate but equal" accommodations for whites and blacks (common among Virginia buildings) not be enforced.[42]

Such was the pace of life in wartime Washington that by April 1942, with parts of the building still under construction, War Department employees began moving in.[43] The next month the War Department announced that the building would be named simply "the Pentagon" after its shape, and in January 1943 construction was completed. There was no elaborate dedication ceremony. One must be on guard against finding historical causes or historical inevitability in events that only make sense in hindsight. The decision to locate the War Department in Virginia would have implications for the later location of defense contractors in Tysons Corner, ten miles west. Obviously that decision was not the only cause, nor is Tysons Corner the only place in the region where defense contracting goes on. Northern Virginia's history was, however, shaped as much by FDR's car ride in August 1941 as it was by the events of 1861 to 1865.

WSEG, IDA, 1946-1960

For the initial, bold plan for a consolidated War Department building located in Virginia, Somervell agreed to design a structure of sufficient floor strength so that, after the war ended, it could be converted into a storage facility for archives.[44] The assumption was that the War Department would shrink back to its 1930s size and move to a proper place back across the river. Somervell probably did not believe that, but there were influential critics of his plan, and he had to reassure them that this enormous building was not going to be an embarrassment and scandal to the government after the hostilities ended. Well before the German and Japanese surrenders of 1945, it was clear that there would be no return to the status quo ante bellum. The Pentagon would have no difficulty remaining filled with workers.

FIGURE 2.1

The Pentagon. *Source*: National Air and Space Museum, Archives, Desind Collection

Vannevar Bush was among those who realized that the much-expanded military, and the alliance between it and the academic community of scientists and engineers, would have to continue in some form. In the 1930s he was critical of the New Deal and its notion of an expanded government role in the American economy. By 1945 he worried that there was nothing to check the appetites of the military, which had made such effective use of powers of nature unleashed by scientists, and which would not want to cede any control over those powers. Bush did not enjoy a close personal relationship with Harry S. Truman, who became president upon Roosevelt's death in 1945. But both men shared a desire to unify the armed services and reduce the competing and overlapping work they intended to do in the post war era. In 1946 Truman took an initial step, based on a proposal by

Bush, which in a linear fashion would lead to the companies inhabiting Tysons Corner forty years later. That was the creation of a Joint Research and Development Board (JRDB): an advisory body, located in the Pentagon and consisting of both military and civilian members, who would "coordinate all research and development activities of joint interest to the War and Navy Departments so that the War and Navy Departments will establish and carry out a strong, unified, integrated and complete research and development program in the field of national defense."[45] The next year (1947), Truman signed a law unifying the services into the "National Military Establishment" (after August 1949 the Department of Defense), with a separate air force. The JRDB was reconstituted as the Research and Development Board (RDB), reporting directly to Secretary of Defense James V. Forrestal. It had essentially the same charter, with two salient features. The first was to coordinate and reduce overlap among the services' weapons programs (especially the development of nuclear-tipped guided missiles). The second was to ensure that civilian research scientists of the highest caliber held positions of authority when decisions about weapons were made.[46]

The RDB was unable to assume the authority that Bush hoped it would and lasted only until 1953. It foundered on the notion that civilians might encroach on the strategic use of weapons in warfare, an authority the Joint Chiefs of Staff would not yield. To telescope a convoluted story, the outcome of discussions between military officers and civilian scientists was the establishment of yet another group, the Weapons Systems Evaluation Group (WSEG), in December 1948.[47] Like its predecessors, the WSEG served all branches of the military and had a combined military and civilian management structure. But its charter was restricted to the evaluation of weapons systems. Perhaps for that reason it was well regarded and lasted longer, until 1976. Its evaluation of weapons systems was to be based on "advanced techniques of scientific analysis and operations research"[48] Reflecting the conflict over civilian scientists in the military, it had two directors. Lieutenant General John E. Hull was its director; its research director was Philip Morse, the proponent of operations research.[49] The group was intended to be modest in size; by September 1949 it employed forty-three people, civilian and military. Some of the staff were borrowed from RAND and from the Operations Research Organization. Finding people qualified in this new field was difficult, and hiring them away from university research positions to come to Washington doubly so.[50] Unlike RAND, the WSEG's offices were close to its sponsors and were located in the Pentagon building. From an operational standpoint it was much more beholden to the immediate needs of the Joint Chiefs of Staff, much to Bush's dismay.[51]

From 1949 to 1955 the WSEG produced at least 16 major reports, which were delivered to the Joint Chiefs (table 2.1). They reflected the tensions of the Cold War with the Soviet Union: reports on chemical, biological, radiological, and atomic warfare; air defense; deployment of submarines and large aircraft carriers; and defense against Soviet subs. Clearly the WSEG was a place where people thought a lot about the unthinkable: an exchange of nuclear weapons with the Soviet Union, which exploded its first atomic bomb in August 1949. The first of those studies was in response to a heated debate between the Navy and the Air Force over funding for the construction of large aircraft carriers versus heavy bombers, especially the B-36. That study addressed a personal memorandum from President Truman to the Secretary of Defense, which raised two questions: could the Air Force, if asked, deliver atomic bombs to desired targets within the Soviet Union? And, if delivered, what would be their effects?[52] As the interservice dispute over the B-36 had spilled out of the inner halls of the Pentagon and was becoming an embarrassment to the President and Congress, this first study was guaranteed to legitimize the WSEG and justify its existence.[53]

By 1955 it was clear that the group was failing in its mission. The principal cause was one foreseen by Professor Morse, namely the inability to attract top talent to the WSEG, which competed directly with RAND and other organizations that could pay higher salaries, had higher prestige, or both. WSEG's reports were of high quality, but too few in number, and it never built up a staff of qualified people to satisfy the requests of their sponsor. Just as the WSEG was founded in response to failings of the Research and Development Board, so too was the response to WSEG's problems the founding of a new organization, the Institute for Defense Analyses (IDA). Employees of this institute would assist the WSEG in responding to requests for unbiased, scientifically based advice. IDA employees would not be civil servants, but could be paid higher salaries and given a measure of independence not available to those on the federal payroll. In short, IDA would have the prestige and quality of RAND, but because of its location in the Pentagon, less independence. Its location also made it difficult for IDA to chart an independent course from the WSEG. In 1964 IDA moved out of

TABLE 2.1
WSEG Reports, 1949–1955

Report No.	Subject	Date
1	Effectiveness of strategic air operations	Feb. 8, 1950
2	Toxic chemical agents	July 11, 1950
3	Artillery-delivered atomic weapons	July 25, 1950
4	Air defense weapons and weapons systems (first interim report)	Dec. 27, 1950
5	Allied capabilities to carry out the ocean transport requirements of current emergency war plans in the face of estimated Soviet submarine and mine threats (First interim report)	June 29, 1951
6	Military capabilities of the nuclear-powered submarine (First interim report)	Dec. 10, 1951
7	Offensive and defensive capabilities of fast carrier task forces in 1951	Feb. 20, 1951
8	Offensive biological weapons systems employing manned aircraft	July 15, 1952
9	U.S. capabilities in 1956 and 1960 for employment of radiological warfare systems in air and ground operations	Aug. 26, 1953
10	Effects of the mid-1954 first phase atomic offensive against fixed industrial targets in the Soviet bloc	Oct. 14, 1953
11	Military worth and effectiveness of ground force weapons systems with air support and atomic weapons	Mar. 22, 1955
12	Combined U.S. atomic offensives in a war beginning in mid-1955 (summary report)	Feb. 28, 1955
14*	The status of biological warfare weapons systems	June 1, 1955
15	Continental defense	July 8, 1955
16	Air interdiction of ground logistics	Aug. 19, 1955

Note: * No report 13 was listed; it may be classified.
Source: John Ponturo, "Analytical Support for the Joint Chiefs of Staff: The WSEG Experience, 1948–1976," Arlington, VA: Institute for Defense Analyses, 1979, p. 101.

the Pentagon, but only to 400 Army-Navy Drive, across the street. Today it occupies a building in Alexandria, still close by.[54]

Like RAND and a few other federal laboratories established after World War II, IDA could pay higher salaries because of its status as a Federally-Funded Research and Development Center. This device, known by the intimidating but, to Washington insiders, familiar, acronym FFRDC, was one of the most significant managerial innovations of the postwar era; an innovation in government that was as significant as Vannevar Bush's innovations early in the second World War. The establishment of the FFRDC had an impact on hiring practices in the Washington D.C. area and affected the conduct of federally funded research throughout the country. It has been the subject of a shelfful of studies and has been attacked on one side as corroding the civil service ethic of service to one's country, and on the other as short-circuiting the open system of bidding for contracts by for-profit corporations.[55] Tensions run high between the for-profits and these centers—especially the MITRE Corporation, which has had a suite of offices in Tysons Corner since the mid-1960s. Yet they coexist side-by-side, even sometimes as tenants in the same office buildings around northern Virginia.

Initially IDA was sponsored by a consortium of universities: Cal Tech, Case, Stanford, Tulane, and MIT.[56] James R. Killian, president of MIT, was the chairman of its board. Albert G. Hill, of MIT's Lincoln Laboratory, was chosen as its director of research. Like its predecessor agency, the IDA focused on studies of war with the Soviet Union using nuclear weapons, the military effectiveness of ballistic missile programs then under development, and whether defense against such missiles was feasible. For the rest of the decade the IDA produced dozens of major reports on these and related topics.[57]

INTERLUDE: GEOGRAPHY IN WASHINGTON, 1950–1957

At the time of Sputnik's launch in October 1957, the descendants of the World War II operations research groups were concentrated in or around the Pentagon, with a few modest facilities scattered nearby. The dairy farms around Tysons Corner were giving way to suburban housing, but commercial development there consisted of modest suppliers of building materials, construction companies, automobile dealers, and others who served the real estate development. The landscape to the west, along what is now the Dulles

Corridor, was still rural and agricultural. One of the few industries there was the Bowman distillery, where Virginia Gentleman bourbon was produced. It was founded in 1935, after repeal of the Eighteenth Amendment, along the W&OD Railroad.[58] Across the Potomac, the Maryland suburbs, already full of bedroom communities, were also beginning to receive some of the laboratories and research facilities that initially had been in the District.

Imagining a nuclear exchange with the Soviet Union was on the minds of many in Washington besides those in IDA and WSEG. The city, after all, was the seat of government and therefore a prime target for Soviet bombers, assuming they could penetrate U.S. air defenses. American physicists who had visited Hiroshima and Nagasaki in late 1945 knew what a single atomic bomb could do to Washington. The result was one more factor leading to dispersal and suburban growth, along with rising automobile ownership, the decline of streetcar transportation, lending policies that favored single-family homes, and the like. For Americans in Washington and other cities, moving to the suburbs was not only a way to get some fresh air, it was also a way to get beyond the blast range of a Soviet atomic bomb. During a brief interlude, from about 1946 to 1954, dispersal as a means of surviving an atomic attack played a significant role in shaping federal housing policies, although it is difficult to extract that factor from the others mentioned. Rapidly deteriorating relations with the Soviet Union, especially after the Berlin crisis in 1948, accelerated the process. Issues of urban dispersal were discussed by physicists shortly after the Hiroshima and Nagasaki explosions. They understood before anyone else that the real secret of the atomic bomb was simply that it was possible, a secret the United States revealed in August 1945. The Soviets proved that point when they exploded a bomb in 1949, although recent research shows that the Soviet accomplishment relied on espionage.[59] In a series of provocative books and unclassified articles in semipopular journals including the *Bulletin of the Atomic Scientists,* physicists argued for dispersal of industries, construction of highways for evacuation routes, even replacing traditional cities with long linear corridors of housing, factories, and offices.[60] Politicians recognized that Americans were not going to give up their desire to live where they wanted to based on the directive of a remote federal agency, but the debate went on. It died out a few years later after the development, by the United States and Soviet Union, of the hydrogen bomb. That bomb's power made it clear that in an exchange of such weapons, few people would survive no matter where they lived.

After World War II, control of atomic weapons was transferred from the Manhattan Project to the newly established Atomic Energy Commission, with offices on Constitution Avenue, NW. Following the news of the Soviet's bomb in 1949—an event many in the AEC were unprepared for—commissioners worried that they would be unable to carry out their duties in the event of an atomic attack. The Commission relocated itself to Germantown, Maryland, twenty miles from downtown and as late as 1985 still surrounded by farmland. There it moved into a specially designed, blast-resistant building. The twenty-mile distance was based on the calculated effects of a twenty-megaton bomb, which was felt to be the largest the Soviets could possibly develop.[61] The major axis of the building was north-south, to further protect it from blast damage. The AEC's relocation was but one of many: fear of an atomic attack led President Truman to prohibit the construction of major new government facilities in the District in general, an edict that affected scientific and technical facilities.[62]

Although the War Department did not look at the Maryland suburbs for its new headquarters building in 1940, those suburbs did become the home of many federal research and military laboratories after World War II. The first move actually occurred before Pearl Harbor, when in 1940 the National Institute of Health moved to a site in Bethesda. In 1946 the Naval Ordnance Laboratory moved from the District to a site at White Oak, northeast of Silver Spring. It was the home of advanced naval research and remained there until 1977, when it was closed as a result of Congress' Base Realignment and Closure (BRAC) actions. During World War II, two sites emerged where the activities of signals intelligence, cryptanalysis, and communications security were concentrated: the Navy on Nebraska Avenue in the District, and the Army at Arlington Hall in Virginia. (Both had been girls' schools before World War II). In 1950 the Joint Chiefs decided to consolidate these and similar activities, and in 1952 they chose Fort Meade, Maryland, about twenty miles from downtown Washington and halfway to Baltimore. Fort Meade was close to Washington and would not require the highly skilled and well-educated work force to move. It was underutilized as a fort, with a large tract of land used to feed and shelter draft animals that pulled Army cannon. That gave the site good security for an operation that had to be conducted in secrecy.[63] The initial building was completed in 1957. The consolidation was called the National Security Agency, a name that itself was secret for a while. The Army and Navy retained facilities in

Arlington and the District, respectively, but most were relocated to Maryland.

In March 1942 the OSRD established a laboratory in a garage in downtown Silver Spring, Maryland, just over the District line, where the proximity fuze was developed under the direction of Merle Tuve of the Carnegie Institution. By exploding near enemy aircraft without necessarily making a direct hit, the fuze dramatically increased the effectiveness of Allied anti-aircraft weapons. At the war's end it was judged on par with the atomic bomb and radar as among the technical advances that ensured an Allied victory.[64] The lab, called the Applied Physics Laboratory and administered by Johns Hopkins University, moved to Laurel, Maryland and continued to develop advanced weapons systems, as well as space-based military and civilian systems.

The trend of moving federal laboratories to Maryland continued into the 1960s. In 1965, the National Bureau of Standards moved to Gaithersburg, not far from the AEC site in Germantown. Renamed the National Institute of Standards and Technology (NIST) in 1988, it remains one of the premier federal labs.[65] NASA's Goddard Space Flight Center, initially called the Beltsville Space Center, was established in 1959, and in 1961 it moved into a building on federal land in Greenbelt. The staff there came initially from several District sites, especially from the Naval Research Lab on the Anacostia River, the Space Computing Center on Pennsylvania Avenue, and a lab in Silver Spring at the District border.[66] The Central Intelligence Agency was an exception to this Maryland rule: it relocated from the District to Langley, Virginia, between 1960 and 1963.

Thus by 1965 a pattern emerged. The District and Maryland suburbs were home to numerous laboratories operating under civil service rules and engaged not only in research, but also in the development of weapons or space hardware. Virginia became home to FFRDCs, which, like RAND, performed analytical studies but left the development and manufacture of hardware to others.

From the perspective of the twenty first century, one finds these discussions about the dispersal of technical facilities ironic. There was no atomic attack on Washington or any other U.S. or Soviet city, although both nations came close at times, especially during the Cuban Missile Crisis of 1962. Cities are vulnerable to terrorist attack, but their health is also seen as vital to the country today, and the terrorists attacks of 2001 did not lead to a movement to abandon New York or Washington. Consider also the arguments for dispersal in the context of discussions about Silicon Valley. The students of that region, and those who wish to replicate its success elsewhere, nearly all emphasize the intense concentration of talent in a relatively small area—not as dense as Manhattan or San Francisco, but not dispersed across the countryside, either. Garreau's analysis of edge cities is similar: there are benefits derived from locating information services and other high-technology activities in a compact, if suburban, area. We shall see that Tysons Corner reflects this paradox: its origins owe a lot to a desire to leave Washington, D.C., while its current prosperity derives from its concentration of talent and technical resources.[67]

A final postwar development brings us back to the transformation of Tysons Corner into a center of defense contracting. Deteriorating relations with the Soviets in 1948 led the Joint Chiefs of Staff to ponder yet another unthinkable question; namely, how to ensure the continuity of government in the event of a nuclear attack on the city. The Defense Department began building or adapting facilities in hardened underground sites, located in a "Federal Relocation Arc" in the hills and along the Blue Ridge, the first range of the Appalachians, ranging from 40 to about 200 miles from Washington. One facility was located under Mount Weather, named after the U.S. Weather Bureau, which established a station there many years before. Another was along the Blue Ridge just over the Mason-Dixon Line in Pennsylvania, another near Warrenton, in the Piedmont region of Virginia, and another deep under the Greenbriar Hotel in West Virginia. There may be others whose existence has never been revealed. When news of the Greenbriar's secret role was exposed in the mid-1990s, it caused some chuckles and ridicule, but after September 11, 2001 those decisions are looked on in a more sober light.[68]

The Pennsylvania site, on Raven Rock Mountain, was designated the Alternate Joint Communications Center (AJCC) and was established to preserve lines of command from the President and military leaders to military bases and troops in the field.[69] A radio transmitting station in Greencastle, Pennsylvania and a corresponding receiving station in Sharpsburg, Maryland, both on the other side of the Blue Ridge, were built to take over in the event of an attack on the six transmitting and receiving stations in the Washington area, two each for the army, navy, and air force.[70] A tower just south of Raven Rock relayed signals from these two backup sites to

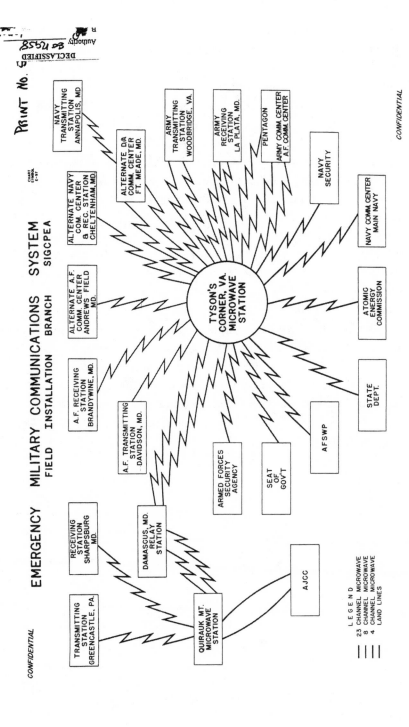

FIGURE 2.2

Emergency military communications system for the capital region. The army implemented this plan for emergency communications in the early 1950s. Although the Tysons Corner radio tower is central to this plan, for most of its existence only a modest staff has ever worked in the small building at the base of the microwave tower. *Source:* National Archives the six stations mentioned above, the Pentagon, White House, the State Department, and other facilities in the Washington D.C. area. Microwave signals would not carry directly from the Blue Ridge to Washington, so intermediate relay towers were needed. These had to be close to the Pentagon and White House, but it was imperative that they be located far enough away so as to be safe from atomic attack on the city itself. Tysons Corner, recognized during the Civil War as a favorable site for signaling, was ideal. The Army built a microwave tower there in 1952 that served as the major relay from all the Washington defense communications facilities to their alternates to the west. It is still in use. The equipment installed in 1952 is as obsolete as the semaphores used by Union scouts at Peach Grove in 1862, but the military still controls the towers. That is all one can say about the tower's use today. It is probably not used for numbers transmissions, but conspiracy buffs have not ruled out that possibility.[71] For many years the tower dominated the Tysons skyline, although recent high-rise office construction has obscured its prominence. Direct photography of the tower is still forbidden, however, and those who are tempted to take a picture for a souvenir run afoul of the Internal Security Act of 1950, as a weather-beaten sign at its base informs us. The numerous defense contractors who took up residence there in the following decades did not do so because of the Corner's elevation or unimpeded view of the Blue Ridge, but for other reasons. Nevertheless, the tower's presence serves as a useful reminder of the military and defense roots of Tysons Corner's economy.

If the Washington suburbs were superimposed on the hands of a clock, the Dulles Corridor would extend out from the ten o'clock position. Across the river, at two o'clock and an equal distance from the White House, is the town of Greenbelt, Maryland. Franklin D. Roosevelt played a role in its creation, too: in 1934 he took a drive with Rexford Guy Tugwell, his undersecretary of agriculture, out to the Agricultural Research Center. Tugwell suggested that a new town be built there for the settlement of low-income families and for those employed at the agricultural center and other agencies. FDR liked the location, and the next year, under the auspices of the Resettlement Administration, ground was broken for the new town.[1] By 1937 the town had taken on its distinctive name, and the first settlers began moving in.

By almost any measure, Greenbelt could hardly have been more different from what developed on the Virginia side. It was the product of the most pure New Deal philosophy: that the federal government could intervene in the private business of housing development for the benefit of those affected by the Depression. The town was planned in line with the bold theories of British planner Ebeneezer Howard, who proposed "garden cities" as an alternative to big city development.[2] The town was to be built in harmony with the shape of the land, not by overlaying a grid of streets without regard for topography. Central to Howard's theory was the "green belt" itself, for which the town was named: a buffer of undeveloped land that would surround the town and on which construction would never occur.[3] Growth, when and if it did occur, would take place in new, similarly-planned garden cities, with their own green belts. Howard's ideas resonated at the time with a number of urban planners and architects, including Lewis Mumford and Frank Lloyd Wright, whose 1930s plan for a "Broadacre City" shared some of its qualities. Both FDR and Eleanor Roosevelt followed the town's early years closely. The First Lady visited at least once, shortly after Pearl Harbor, and to this day residents of the town regard her as one of the guiding forces behind its vitality. It is no coincidence that they named the local high school, one of the best in Prince Georges County, after her.

By the late 1940s political winds were blowing in other directions, and what were once matters of pride—a visit by Eleanor Roosevelt, a government-sponsored test of ideas about planned communities—became liabilities. The town contributed its share of defense workers and soldiers to the Second World War, but its close association with the New Deal made it a target for those who wished to roll back the changes FDR had made two decades before. Greenbelt's food-purchasing co-op was denounced as "communistic." Government workers found themselves under suspicion simply because they were residents; some were dismissed from their jobs outright.[4] In 1953, five out of thirteen residents Greenbelt who worked at a division of the Navy were identified as security risks and suspended from their jobs. The five suspended were all Jewish; the remaining eight were not.[5]

Among those dismissed was Isadore ("Izzy") Parker, a draftsman whose alleged offenses including associating with Frank Lloyd Wright (who was hardly a communist but was an outspoken man with a reputation as a free-thinker). Parker was perhaps too well known for his cartoons that appeared in the town newspaper—some of which gently poked fun at an establishment that, in the 1950s, had lost its sense of humor. In Parker's words, the whole affair was a matter of nothing more than "guilt by association."[6]

Parker's architectural drafting talents were in demand in the booming 1950s, however, and in 1955 he got a good job with the private engineering firm of Michael Baker, Jr. There he would play a key role in the detailed design of what would be, like the Pentagon, one of the region's most ambitious and influential construction projects: the Washington Circumferential Highway, later named the Capital Beltway.[7]

THE CAPITAL BELTWAY

It is no surprise that the Atomic Energy Commission wanted to locate its facilities outside Washington's atomic bomb blast zone. It may surprise us, however (especially those who commute on the Beltway every day), that this highway's location also reflects worries over atomic blasts. Such was the case with the road around the Capital, and indeed with the Interstate Highway

System as a whole. The highway that has ringed the city since 1964 was the outermost of three ring roads planners had envisioned. A plan adopted by the National Capital Planning Commission in 1950 proposed two inner loops, one about a half mile from the White House, another at about a six-mile radius, and finally an outer loop, at a ten-mile radius. This outer loop became the Beltway.[8] (A fourth loop, even farther out, was mentioned but not described in detail; a few segments were eventually built as the Fairfax County Parkway in Virginia and I-370 in Maryland.)

Fragments of the innermost loop were built before being stopped by local opposition. These fragments became parts of Interstate 395, which runs south of the Mall through a tunnel under the Reflecting Pool in front of the Capitol, and an interchange connecting Interstate 66 to local streets in front of the Kennedy Center. The intermediate loop was less well defined, and was at times conflated with a pre-World War II plan to build a parkway to connect the ruins of the Civil War forts that defended Washington. Only a small segment, now Military Road near Fort Stevens, was built. Plans for this loop were intertwined with plans for a complex skein of freeways going in and out of the city, mainly from the north along Sligo Creek, Wisconsin Avenue, the B&O Railroad tracks, and through other corridors. Few of these were built, again due to opposition from the affluent communities through which they were to run. One exception was Interstate 66, built into the city from the west. That was built years later, and its construction was so unique that I will discuss it separately in chapter 6.[9] Another exception to this pattern was a highway built to relieve congestion on Route 1 in Virginia, coming from the southwest. Named after Virginia Highway Commissioner Henry G. Shirley, the first segment opened in 1944 and was tied to the newly opened Pentagon. Despite wartime gasoline and rubber rationing, the Pentagon was and remains a building best served by automobile, and about $18 million was spent to build highways to access it.[10]

That leaves the so-called outer loop, envisioned just beyond the ten-mile square borders of the District. Early plans were refined by 1952 and were supported in Congress by Senator Francis Case of South Dakota (after whom a bridge over the Washington Harbor is named). Defense issues were never far from these discussions. One description mentioned how a bypass loop would give "tanks from Fort Belvoir, say, a route north to cut off an aggressor force." Francis C. Turner, whose later position as the federal highway administrator gave him great influence over the Interstate Highway System, discussed the advantages of a having loop in addition to the highways penetrating the city by saying, "So in case a bomb dropped, like Hiroshima, the military needed a route to go around the city, to bypass it."[11]

In 1956 the Interstate Highway Act became law, and plans for a Washington outer loop were subsumed into the plan for a 41,000-mile network of limited-access highways for the nation, with the federal government assuming 90 percent of the cost. The initial impetus for the act came from several senators and congressmen, including Senator Albert A. Gore, a Democrat from Tennessee and the father of the senator who championed an information superhighway thirty years later. Gore took the bold step of introducing legislation to authorize a plan that would be financed by congressional appropriations. President Eisenhower also wanted the roads, but was wary about enlarging the federal budget. The compromise that enabled the act to pass was to have the roads financed by a Highway Trust Fund, paid for by taxes on gasoline, and thus insulated from the general federal cycle of budgeting, appropriations, and taxes.[12] Despite his opposition to Gore's plan, Ike's personal interest in an interstate system was crucial to its creation. As a general, he had ridden on the German Autobahns in 1945 after V-E Day, and he recognized their military value. As President, he appointed his wartime colleague General Lucius Clay to lead the effort to get a highway bill passed.[13] Eisenhower remembered how the Allied bombing of the German rail network was much more effective than the bombing of German highways. That was due to the nature of a railroad line, which is more quickly shut down by damaging only a short section.[14] After becoming President, Ike specifically mentioned limited access highways as potential evacuation routes "in case of atomic attack on our key cities." Note that he was thinking of highways that penetrated, not bypassed, the cities. Nor does his statement imply an understanding that new urban centers might appear at the interchanges of these new highways.[15]

After its completion in 1964, the Capital Beltway became so successful as a commercial highway, so congested, and so much a part of the social fabric of Washington that the military's desire to have the road outside an atomic blast zone was forgotten. Fear of a Soviet attack was real, and from time to time increased—for example, during the Cuban Missile Crisis in 1962. But few drivers thought of that as they began to use newly opened sections of these highways. Nevertheless, of the four ring roads planned for the nation's capital, only the Beltway was fully completed, a fact that at least one observer attributes to its military importance.[16]

Detailed planning began in March 1954, when representatives from Virginia, Maryland, and the District met in a District hotel and agreed to develop a route. Of critical importance was the decision, made at that meeting, of where to locate the two Potomac River crossings: Jones Point for the lower, and Cabin John, Maryland for the upper crossing. At this stage, two years before the formal establishment of the Interstate Highway System with its federal funding formula, the planning was still in the hands of the Virginia and Maryland State highway commissions. The highway that resulted from those initial efforts reflected that fact.

The subsequent history of the design and construction of the Beltway is a fascinating story. In spite of its importance to Washingtonians, few know it in full. Many myths about the Beltway's location and construction persist in spite of several fine scholarly treatments of its history. The Interstate Highway System ranks at or near the top of twentieth-century engineering achievements in the United States, yet because of its dispersed and prosaic nature the public does not perceive it as such.[17] This narrative will look briefly at three facets of the Beltway's history: the location of the upper Potomac crossing at Cabin John; the different natures of the Maryland and Virginia suburbs as they affected the road's location, and finally, an aspect of the Beltway's social impact.

One of the most persistent myths about the Beltway's location is that its route was influenced by wealthy landowners, who either wanted the road to come close to their property for commercial reasons, or did not want it to ruin their backyards or neighborhoods. Road building has always been a political and social as well as an engineering activity, and the Beltway was no exception. The Beltway project exhibited an interesting difference between the pressures exerted in Montgomery County, Maryland, and Fairfax County, Virginia. We have already seen that the initial concept for this road came from political (in other words, Cold War), not engineering, needs. A simple narrative of the process would begin there, proceeding to the engineers' requirements for such a highway. After the initial concept was developed, engineering would take over, placing the highway where it was best from the standpoints of safety, speed, comfort, and cost. As work on that phase progressed, politics would intervene again: a person or influential group would object to plans published in the newspaper or elsewhere, causing the alignment to shift (or not, depending on the person's political clout). This process would be iterated until a final plan was established and the road built.

For the Potomac crossing at Cabin John, this process did occur. The initial location was based on the military need to be away from the center of the city and the traffic engineer's need to provide for a bypass around its congestion. The Potomac at that point runs through a deep gorge, requiring a high and expensive bridge. Engineers chose a narrow spot to minimize the costs. That choice, however, passed over Plummers Island, a wildlife preserve that since 1908 had been operated by the Washington Biologists' Field Club. At the Club's insistence, the bridge was moved about 200 yards upstream, where it just barely crossed the island's northern tip.[18] The crossing, initially called the Cabin John, later renamed the American Legion Bridge, opened in December 1962.

The alignments on either side of the bridge reflected different political forces. To the north, the route passed through Maryland suburbs that were long developed and settled with affluent homeowners. Beginning in the 1880s, developers such as Francis G. Newlands had extended the avenues of L'Enfant's plan out past the District into Maryland. By the 1920s trolley lines were built in the medians or alongside these avenues to the suburbs of Chevy Chase, Kensington, Cabin John, and Bethesda, to Great Falls on the Maryland side, and to the established towns of Rockville, Silver Spring, and Laurel. Like their counterparts in Virginia, most of these lines were abandoned by 1939 (the trolley to Cabin John was the last to go, in 1960).[19] But the pattern of suburbanization remained and intensified, with the avenues providing easy automobile access directly into the heart of the city. The result was that the planners of the Circumferential Highway had to thread a route through an area already settled with wealthy and influential people, as well as through parkland that was dear to the county's residents. The results were compromises that got the highway built, but which in the end were unsatisfactory. In the section between Wisconsin and Connecticut Avenues, the road was connected to an already-built, but narrow parkway running through Rock Creek Park. That avoided having to condemn houses nearby, but as the Beltway expanded, the sinuous and now-dangerous path of the original parkway served as a daily reminder of the original compromise. Worse, each periodic road widening took away more and more parkland.

There were cases of influential residents shifting the road alignment here and there, which added more curves. One persistent legend is that a wealthy Democratic fundraiser caused the highway to be shifted to spare a tree. (I have never found solid evidence to support this story and have been told by one principal that it is not true.) The Beltway does make a slight curve at that point, near the railroad overpass in Kensington. To the west, at the interchange with Rockville Pike, a plaque identifies an oak tree as "the Linden Oak," over 250 years old and one of the largest in the county. That tree is still standing, although in poor health. If other trees closer to the road were spared, they did not last long, as the road has been widened several times since its initial opening and will probably be widened even more in the future. In any event, highway engineers faced enormous pressure from all sides as they sought a good alignment.[20] Democrats and Republicans were affected equally. To the east, the road sliced through and nearly destroyed Greenbelt, home of Izzy Parker and fellow New Dealers. Further west, the Burning Tree Country Club, a favorite golfing spot for Republicans, lay in its path. In between were the modest homes of middle class residents of all backgrounds, who found their neighborhoods broken up, backyards taken, and open land lost forever.

On the Virginia side the conditions were different. The new highway's path was only lightly settled, and rather than suburbs, its planners found wooded valleys and dairy farms. Suburbs had grown along the route of the Great Falls (Virginia) trolley line, but beyond McLean these were sparse. The same was true of development along the major highways, such as Lee Highway and Route 50, extending west from Arlington. By the late 1950s some of the farms along these roads had been sold to developers for housing, but the developers had advance warning that the highway was coming through, and they held back construction in the road's corridor. Among the oldest suburbs was Pimmit Hills, begun in 1950 and now just inside the Beltway. It was intended from the start as low-cost housing for returning World War II veterans, and for many years the residents of nearby McLean looked down on its cheap houses.[21] In his interviews with Joel Garreau, the developer Til Hazel recalled his surprise at the success of Pimmit Hills, and how it revealed to him the future of Fairfax County.[22] Like other inexpensive postwar tracts, Pimmit Hills has grown up, with the original saplings now mature shade trees, and many of the houses added to and lovingly cared for by their owners. The houses there are no longer cheap.

The Virginia highway planners found a swath of open land just beyond Arlington and McLean, but not so far out that it ran into the existing towns of Vienna and Fairfax City. They had a much easier task than their Maryland counterparts: from the bridge at Cabin John the road followed a straight line through the valley of Scott Run, which conveniently ran in a northeasterly direction to the Potomac. The valley was swampy but undeveloped. (How times have changed: today swamps are called wetlands, and highways are routed around, not through them.) Leaving that valley near Tysons Corner, the alignment created a triangle with the existing routes 123 and 7 (the old Chain Bridge Road and Leesburg Pike, respectively). As far as anyone can recall, the creation of that triangle was a coincidence.

From Tysons Corner, draftsmen drew a set of broad, sweeping curves that turned the road gently to the south, then again to the east, crossing the Potomac River at Jones Point. Whereas the Montgomery County planners had to condemn and purchase homes and other properties by eminent domain, their counterparts in Fairfax County could negotiate with local farmers and landowners who were willing to sell, recognizing that the highway would increase the value of their remaining holdings. Surveyors who walked the Maryland route found themselves walking through golf courses and near the back yards of members of Congress and their staffs; surveyors in Virginia saw only "scattered homes ... many had outdoor plumbing rather than indoor plumbing, and the kids walking barefoot."[23]

Sections of the Virginia highway opened in 1961; by April 1964 all twenty two miles in Virginia were open. On August 17, 1964, the entire loop was dedicated at a ceremony held at an interchange in Maryland. Its total cost has been estimated at $189 million in 1960s dollars, which one may compare to the Pentagon's cost (including roads) of about $54 million twenty years earlier. By 1960 the awkward name "Washington Circumferential Highway" had been discarded in favor of the more memorable "Capital Beltway," although initially there was confusion of whether it was to be named after the Capitol (the building where Congress meets) or the capital (the seat of government). The latter spelling prevailed, and Capital Beltway it is.[24]

Once opened, the highway became the catalyst for enormous changes in the social and economic fabric of Washington. These changes occurred all over the United States as the Interstate System was built. Before turning to its effect on Tysons Corner, it is worth mentioning a perhaps minor social effect, which figures in our story in an interesting way. As described above,

the highway bisected and split Maryland neighborhoods and replaced
wooded backyards with the noise and fumes of traffic. Like any new technology, it also attracted criminals who took advantage of its convenience and
speed before police responded with new law enforcement techniques. There
were no noise barriers in those days, and thus the road lay against the back
yards of suburban houses. Thieves soon learned that they could park their
trucks alongside the road, break into an adjoining house through its back
door, and make off with the loot before anyone had a chance to notice. By
using the Beltway as an escape route, neighbors could not see who the
thieves were or what kind of vehicle they were driving. And once on the
Beltway, they could move quickly from one jurisdiction to another, crossing
between Maryland and Virginia in minutes—a journey that used to take
over an hour. They also found it easy to fence the loot in the District, whose
police force had little knowledge of the then-remote suburbs. An especially
notorious ring, based in Fairfax County, was eventually caught and associated with thousands of burglaries. A 1968 newspaper account of the ring-
leader's sentencing referred to him as a "Beltway Bandit." Two years later the
problem was still acute, and after some of those convicted were released from
prison, they were caught doing it again.[25] Eventually the various jurisdic-
tions developed coordinating bodies to exchange information and make this
type of crime less common. The 1968 article is the earliest known use of
Beltway Bandit. It has become familiar to Washingtonians and has been
adopted by a Flyball dog team and is the name of a Frank Zappa jazz com-
position, among other uses.[26]

DULLES AIRPORT

At the point where the Beltway leaves the valley of Scott Run, it intersects
another limited access road that also had a profound impact on the region.
That is the road built by the federal government to provide access to Dulles
International Airport, twelve miles west of the Beltway and twenty five miles
from downtown Washington. Washington National (now Reagan National)
Airport opened in 1941 and replaced Hoover Field, where the Pentagon was
built, but it was not long before aviation authorities recognized the need
for another, larger facility. National's nearby location made it popular, but its
runways were short and its location on the Potomac made it impractical to
extend them. Two existing fields were considered for expansion: Baltimore's

Friendship Airport (now Thurgood Marshall-Baltimore Washington International), and Andrews Air Force Base, just south of the city. The congested skies over Baltimore, on the one hand, and the Air Force's resistance on the other, led to these sites being rejected. Thus planning for a new, second airport began shortly after the end of World War II. In 1950 Congress released funds to the Civil Aviation Authority (CAA), the predecessor to today's Federal Aviation Administration (FAA), to purchase land. The CAA surveyed potential sites and narrowed the choices to two: Burke, Virginia, and Pender, Virginia (the site of the Civil War battle of Ox Hill). The agency's surveyors preferred the site at Burke, about twelve miles southwest of the White House, and the CAA bought a thousand acres there, with plans to buy several thousand more. But before it could buy more land, the agency faced opposition from nearby residents, who enlisted the help of the influential Virginia Senator Harry F. Byrd. In 1955 Senate subcommittee hearings were held wherein Byrd expressed his opposition. An attempt to appropriate more money for the Burke site was tucked into a foreign aid appropriations bill at the last minute, but this was thwarted when Senator Leveritt Saltonstall of Massachusetts caught it. Saltonstall phoned Elwood R. "Pete" Quesada, the "Special Assistant to the President [Eisenhower] for Aviation Matters." To Eisenhower, Quesada was to aviation as Lucius Clay was to Interstate Highways: a wartime friend and trusted confidant whom the President could count on to come up with a political as well as practical solution to the problem. Saltonstall asked Quesada to come up with a better site, and promised a congressional appropriation by 1957 if the site was personally recommended by the Office of the President.[27]

In January 1958 the White House announced its choice: a large tract of land north and west of the old crossroads of Chantilly, through which Stonewall Jackson's army marched in September 1862. Unlike Burke, it was sparsely populated and was almost entirely farmland. It was not empty, however: fifteen miles of roads had to be closed, and 250 pieces of property purchased or condemned by eminent domain. Only about half of the 580 buildings demolished were homes; the rest were barns, sheds, and other farm outbuildings. That was for a 9,800-acre tract of land, over twice what had been proposed for Burke. The land acquisition was a testament to Quesada's vision, as he and others recognized the coming jet age (aircraft like the Boeing 707, certified by the FAA in 1958, required longer runways).

Nearly all published accounts call the site Chantilly, but that is a misnomer. In recent years, as Virginian African-Americans rediscover and reclaim their heritage, it has come to light that there was a small village, settled by freed slaves, where the present runways are located. It was called Willard, named after Joseph Edward Willard, a delegate to the Virginia Assembly from 1893 to 1901, and the son of Joseph C. Willard, who owned the famous Willard Hotel in the District. As with land acquisition for the Beltway, the authorities informed the Willard residents in September 1958—with little warning—that their land was to be condemned. The government paid about $500 an acre for the land. Many of the residents relocated to the village of Conklin, a few miles south on Braddock Road.[28] The only vestige of the site is the name of Willard Road, which loops across Route 50 and up to a nonpublic gate of the airport from the south.

The tract of land's shape may seem strange to one looking at a contemporary map, but it was the result of a straightforward process. The first step was to decide on four runways, two running north-south and two east-west. After consulting wind data from a local weather station, the east-west runways were shifted a few degrees to the northwest. Recognizing the onset of the jet age, the planners decided on long runways: 11,500 feet for the north-south runways, about the distance from the Lincoln Memorial to the Capitol, and 10,000 feet for the east-west runways. To that length they added a buffer of 8,000 feet on either end, both for safety and to prevent development from occurring along what might be a noisy path as jets came into use. They also acquired a buffer of 2,000 feet on either side of the centerline of the runways. That set the shape and extent of the property: fifteen square miles, about two-thirds the size of Manhattan Island.[29] To further mitigate the noise, large tracts of fir trees, which muffle noise and do not attract nesting birds, were planted along the perimeter.

Ground was broken in September 1958, and the heavily reinforced runways, built to take the weight of the newest jets and future supersonic transports, were complete by 1960. Symbolic of the transformation of the region, aggregate for the runways was among the last loads handled by the Washington & Old Dominion Railroad, soon to go out of existence.[30] The single-track railroad was in poor condition, and no attempt was made to refurbish it to provide rail access to the terminal. Also symbolic of the revolution in transportation was that the bodies for the "Mobile Lounges", the large buses that took people from the terminal to their planes, were

FIGURE 3.1

Dulles Airport terminal, 1990. The photo was taken on the occasion of the delivery of an U.S. Air Force SR-71 Blackbird aircraft to the Smithsonian Institution's National Air and Space Museum, which was then in the process of building its annex facility on the southeast corner of the property. The Blackbird's final resting place, not far from the contractors and agencies in charge of photo reconnaissance, is appropriate. *Source*: Smithsonian Institution (SI photo 90-3333)

manufactured by the Budd Company of Philadelphia, a long-time builder of railroad passenger cars. While the airport was under construction in 1959, President Eisenhower named it Dulles International Airport after former Secretary of State John Foster Dulles. The terminal building was completed and dedicated on November 17, 1962, and both President Kennedy and former President Eisenhower spoke.

The press accounts of the opening concentrated on the beautiful terminal building designed by Eero Saarinen & Associates and its spectacular setting in the Virginia Piedmont. The terminal was expanded in 1996 in conformance to Saarinen's original plans, and it remains one of the region's most beautiful buildings. The airport remains one of the best designed in the world, although the Mobile Lounges proved unpopular and are scheduled to be replaced by a subway.

At the dedication, the press gave only minor space to other decisions made during the airport's design, but these proved to be as significant as the location itself. The first was made by Quesada himself, in recognition of the great distance from the new airport to Washington. He specified the inclusion of an expressway from the terminal to Route 123 in McLean, with plans to connect to a future interstate highway into the District, crossing the Potomac at an outcropping of rocks known locally as the Three Sisters. This fifteen-mile road had no access points between the airport and Tysons Corner, where it interchanged with the Beltway. From Tysons Corner it went another mile to end at Route 123, which in turn was widened—over local opposition—to handle through traffic. This remained the terminus of the Access Road for twenty years, before the final leg to the District was completed. By that time the Three Sisters Bridge was cancelled, and the highway entered the District at a bridge a mile downstream.

The result was that the highway provided excellent access from the airport to Tysons Corner and the Beltway, but much less favorable access to and from the District. Its lack of local interchanges also meant that the planned community of Reston, lying alongside but with no access to the road, derived little benefit from this enormous undertaking.

TYSONS CORNER, CIRCA 1963

The failure to complete the access road inside the Beltway meant that beginning in 1963, and for many years thereafter, Tysons Corner became the first stop from the airport, with connections to Maryland and the rest of Virginia via the Beltway. Tysons Corner's strategic position was further enhanced by McLean residents, who, after enduring the widening of Route 123 and the relocation of the CIA to nearby Langley in 1962 and 1963, vocally opposed further commercial development between the Beltway and Arlington.[31] For the next twenty years, then, Tysons Corner, almost exactly halfway between the airport and the White House, was not only the first but also the last practical place for commercial activities between Dulles Airport and the District. That consequence of the highway construction, and the twenty-year hiatus until it was extended to the District, gave Tysons Corner its modern identity. Symbolic of this change was another, minor construction project nearby. In 1963 the Virginia highway authorities began building an overpass to replace the at-grade intersection of Routes 7 and 123, the crossing that gave Tysons

Corner its name. During its construction, whatever vestiges that remained of the original corner—the filling stations and general stores—were torn down, and the bridge that replaced it is somewhat sterile. In practical terms today's Tysons Corner is defined not as that crossing but the area bounded by the Dulles Access Road to the north, Old Courthouse Road to the southwest, and Magarity Road to the southeast.[32]

Just as the spread of commercial growth to the east of Tysons Corner was thwarted by McLean residents, growth in other directions was likewise curtailed. Low-density suburban housing provided a green buffer between Tysons and the older town of Vienna. Planners of the town of Reston provided a similar green belt to the west, which included the Wolf Trap Farm for the Performing Arts, a national park. To the north lay rugged topography overlooking the Potomac. Land had been acquired there for an extension of the George Washington Parkway, but after that extension was cancelled, much of the property became parkland. This pattern of airports, roads, interchanges, and green buffers conspired to make Tysons Corner a compact region poised for dense commercial development—making it, in Joel Garreau's term, the foremost edge city of modern suburbia and giving it a density that is lacking in most other comparable suburbs.

Besides having no local interchanges, the Dulles Access Road had other unique features. The first was that a right-of-way was set aside in its median for a rapid transit line to the city. Some accounts say it was to be a monorail, similar to the ones built at Disneyland and for the Seattle World's Fair. Others say that the type of system was unspecified. But a right of way was reserved. This foresight was admirable, but the failure to construct a rapid transit line at the time was a serious error. As of 2006, there is still no rail transit in the Dulles Corridor, although it is planned for a possible completion by 2012. Meanwhile, traffic congestion is past the breaking point. If such a link is ever built, it will cost many times more than what it would have in 1962.

The second design decision was to set aside a right-of-way on the outside of the access road for a future expressway with plenty of local on- and off-ramps. Details of this, too, were unspecified, and it was not until 1984 that the Dulles Toll Road was opened. But when it was completed it enabled the dramatic leap of development from Tysons to Reston, then to Herndon, Sterling, and beyond.[33] For many years passenger traffic at Dulles remained well below expectations. With the new Toll Road and the growth of Reston and Herndon, it surged to handle 20 million passengers by the end of the century, most of that increase coming in the 1990s.

Another decision made during the airport's construction, even less reported at the time, may have been the most important of all. Initially the airport was to have its own sewage treatment plant, which would discharge treated sewage into the Potomac via the Broad Run watershed. That was unacceptable to Washington, D.C. and suburban Maryland, whose drinking water intakes were on the Potomac downstream from Broad Run. The airport's planners proposed a state-of-the-art plant, and they were confident that the treated sewage would not threaten the water supply. Nevertheless, Washington water engineers feared that it would set a dangerous precedent. They knew that the airport would lead to rapid growth in northern Virginia. Other developments, less well funded or professionally managed, might build plants that were not as reliable or safe. The dispute led to a decision to pipe the sewage to the Potomac, where a sewer line on the opposite bank would intercept the effluent and prevent it from being discharged into the river. This pipe carried the sewage to the Blue Plains Treatment Plant in Washington, below the freshwater intake pipes. This "Potomac Interceptor" was built with future development in mind; Dulles sewage only needed 10 percent of its capacity.[34] As much as the locating of the airport itself, the Potomac Interceptor made possible the rapid development of the Dulles Corridor beyond Tysons Corner. As a side effect, sewer gases vented from the Interceptor make hiking along the C&O Canal on the Maryland side most unpleasant, as new housing developments in Virginia fill the Interceptor up to its capacity. This has led to disputes between Marylanders and Virginians that echo the battles of the Civil War. A dispute over Virginia's right to take water from the Potomac went all the way to the Supreme Court, which in 2003 ruled in Virginia's favor.

RESTON

The opening of Dulles Airport left the new town of Reston, a few miles east, out in the cold. In 1962 Reston was brand new, too—a planned community, like Greenbelt. Its location next to the unplanned edge city Tysons Corner makes it an interesting case study for urban planners. It was planned as a 1960s version of what Greenbelt was in the 1930s and what Washington, D.C. was at the end of the eighteenth century. Not only that, it was built on the site of another utopian town, which was planned in the nineteenth century but never saw fruition. In 1886, General William McKee Dunn, one of the planners of Dunn-Loring, and Dr. Carl Adolph Max Wiehle purchased over six thousand acres along the W&OD railroad line. Wiehle, a German immigrant, hired a city planner from Germany, and together they laid out an elaborate plan for the town of Wiehle, with a symmetrical pattern of streets, three lakes, a post office, and a town hall.[3] Wiehle built a summer home for himself there, along with a schoolhouse, a brick kiln and sawmill, and a hotel. The hotel was popular as a resort, but few lots were sold. The town was abandoned after Wiehle's death in 1901. The location was home to a successful dairy farm, and then the Bowman distillery after the Twenty First Amendment was passed in 1933. The Bowman family took over most of the site and established an estate they called Sunset Hills. They sought more residential and commercial development for their land after World War II, but were stymied by several factors, including the problem of sewage disposal, which was not solved until the construction of the Potomac Interceptor. Thus it was no surprise that the Dulles Access Road had no local exit near Bowman's property.[36] The Bowman family sold its holdings to Lefcourt Realty, who in 1960 resold it to Robert E. Simon, the son of a New York real estate developer. The distillery remained until 1988.

It is not clear how much Simon knew of the ill-fated attempts to build a planned community at that site. In any event he proceeded with plans that exceeded those of the previous owners. Simon was a native of New York and the son of a wealthy and successful broker of properties in midtown Manhattan. Growing up in that environment, he enjoyed the best of what urban life had to offer. Visitors to midtown Manhattan today have to be reminded that in the 1940s and 50s, the area around 57th Street housed not only Carnegie Hall (owned by Simon's father) and the famous restaurants that catered to its wealthy patrons, but also a score of small music shops, rehearsal spaces, inexpensive apartments for musicians and artists, delis, and cheap restaurants to cater to the locals—in short, a diverse and cultured mix of activity. What it lacked was open and green space—Central Park was so beloved because it offered the only respite. At the age of twenty one Simon inherited his father's business and set out to combine the cultural and economic opportunities of New York with the pleasure of living in the country. That led him to the Sunset Hills property.

Before the Second World War, Frank Lloyd Wright and Lewis Mumford wrote about how the new inventions of the automobile and the airplane would free people from the crowded cities demanded by the older technologies of the railroad and streetcar. In an interview conducted many years later, Simon noted that however great an architect he was, Frank Lloyd Wright was a poor urban planner, and Simon mentioned Wright's plan for Broadacre City only as an example of Wright's impracticality.[37] Nevertheless, Broadacre City, conceived in the 1920s and developed at Wright's Taliesin settlements in 1934, was very much in the spirit of Reston's design.[38] Although by 1960 a backlash against automobiles was beginning, most people welcomed, not feared, the auto's ascendancy over rail transport. From the vantage point of today's overcrowded highways, it is hard to remember how much Wright based his ideas on the universal ownership of personal automobiles. His prescription for Broadacre City reflected many of his best ideas, which are as relevant today as ever. It also reveals a reaction against the streetcar and other forms of fixed mass transit, just as Lewis Mumford dismissed steam-powered trains as part of a "paleotechnic" age best left behind.[39] Wright's plan called for "an acre of ground minimum for each individual." To give a sense of how Simon was willing to depart from Wright, one of the first buildings he constructed in Reston was a high-rise apartment building—deliberately planned to let people know he was serious about Reston being more than just another typical suburb. In other respects Reston was like Broadacre City: "No grade crossings . . . No railroads . . . No streetcars . . . No poles . . . No wires in sight . . . No traffic problem . . . No major or minor axis (i.e. no rectangular grid of streets)."[40] City government would not be in the hands of party bosses cutting deals in smoke-filled rooms, but rather there would be "Administration by radio and flight."[41] (For a modern interpretation, change the word "radio" to "e-mail.") Wright's individual houses and public buildings were already famous and revered by the time of his death in 1959; Reston would be a place where Simon could put Wright's ideas for a whole town into practice.

Robert E. Simon drew up a plan, which he presented to Fairfax County for approval, for a town named Reston after his initials. It called for houses, apartments, and commercial buildings separated by trees and green space. Like 1930s-era Greenbelt, a green belt on the east side would protect Reston from sprawl coming out of the District (and Tysons Corner). No equivalent buffer was planned on the other side, and now Reston has

FIGURE 3.2

A publicity photo of Reston, Virginia, mid-1960s. The high-rise apartment building was one of the first structures to be built, reflecting Robert E. Simon's desire to show that Reston was not going to be just another suburb. *Inset*: a sign at an entrance to Reston, announcing its then-unique quality of being a place to both work and live. *Source*: Library of Congress effectively merged with the unplanned growth of Herndon and Sterling. Reston incorporated personal automobile ownership into the design to allow a generous amount of green space around individual residences, businesses, and commercial establishments. That allowed Restonians to have the access to shopping, businesses, places of employment, culture, and their neighbors that New Yorkers enjoyed, but without the high density required for a city that relied on pedestrian traffic. Pedestrians were not ignored, however: they could access the dispersed shops and town centers by using underpasses to avoid mixing with automobile traffic. A reliance on the automobile also allowed Simon to plan for parks throughout the town—ten acres for every 1,000 residents.[42] Above all, by incorporating office and commercial space into the plan, people could both live and work in Reston and avoid lengthy and frustrating commutes, which by 1960 were already recognized as the bane of suburban living.[43]

Simon's plan was bold, and a major departure from what was then the norm in Virginia real estate development. The county approved his plan in 1962, and within short two years residents began moving in. The fulfillment of that plan was not easy, however. The lack of ramps on or off the Dulles Access Road kept Reston from having a good connection to the east, while a planned Outer Beltway to give equivalent access to the north and south was not built until many years later. Meanwhile, traffic at Dulles failed to meet expectations, and that also stifled Reston's growth. Dulles Airport's planners were correct in envisioning a need for long runways to accommodate jets, but they did not foresee the advent of smaller jets, such as the Boeing 727 and Douglas DC-9, which could use the shorter runways at National Airport. As both airports fell under federal jurisdiction, it was up to Congress to decide how to deal with these advances in technology. Around 1964 it allowed the smaller jets to operate out of National, a decision that kept passenger loadings at Dulles lower than anticipated. But complaints about jet noise, especially from residents who lived under flight paths along the upper Potomac, led to an effective cap on traffic out of National by 1981, thus, belatedly, opening the way for growth at Dulles—and Reston with it.

The Washington and Old Dominion Railroad also ran through Reston, but as with the construction of Dulles Airport, the railroad supplied only modest carloads of materials for construction and was soon shut down. Nor was it ever seriously considered for mass transit to Reston.[44] People were moving in and enjoying Reston's open spaces, but most were commuting elsewhere to work by private automobile. The town ran into financial difficulties, and in 1967 Simon relinquished control to Gulf Oil, which bought an interest in Reston at a time when it and other oil companies were diversifying. Gulf sold its holdings to Mobil in 1978, and twenty years later ExxonMobil divested its real estate holdings. In spite of the turmoil, which grew even worse during the 1980s Savings and Loan crisis, the town's development continued, following Mr. Simon's vision, even if not as closely as he would have preferred. As of this writing, he lives in an apartment in town.

Gulf Oil financed a building for the U.S. Geological Survey (USGS), which moved its headquarters to Reston, and which at its peak employed about 3,000 skilled workers. To the disappointment of the town's developers, the USGS did not attract the contractors and consultants that sprouted up around the Pentagon, so it remained an anomaly. In those days Reston's chief attraction was its cheap rents, and it managed to lease space for the back offices—where payroll, accounting, and the like were done—for trade associations and lobbying groups. These groups nevertheless kept their headquarters, where the highest-paid employees worked, in the District.[45] A few high-technology companies were persuaded to move to Reston, beginning with the Silicon Valley computer company Tandem, which opened an office there in 1977. A crucial factor in that decision was a presentation by officials from Reston and from the Dulles Airport Authority showing that Tandem's executives had as convenient access from Dulles to both San Francisco and Germany (its European base) as they had through other East Coast airports. Another early tenant was Sperry Univac (now Unisys), which moved there in the late 1970s.

Reston remained a wonderful place to live for those whose lives centered around access to a personal automobile. For the lucky few who both lived and worked in the town, it was the fulfillment of Simon's vision. Reston may be the closest realization of Wright's ideas for a Broadacre City, although Wright died a few years before people began moving into those towns. There is no indication that the strong-willed Wright would have approved of Reston (or of Columbia, a similar planned town built at the same time in Maryland). Simon's genius was in understanding that such a vision means nothing if it cannot be sold to conservative groups like bankers, construction companies, and politicians. In any event, Reston's streets were attractive and designed with enough capacity to make any part of the town accessible by car in a few minutes. High-paying jobs remained scarce until the impasse was broken by the construction of the Dulles Toll Road in 1985, followed by the Fairfax County Parkway a few years later. Quesada set aside room for a commercial road when he planned the Dulles Access Road, but the issue of financing led to a twenty-year delay in building that highway. Charging tolls was the solution, and that, coupled with an increase in defense spending in the 1980s, transformed the once-sleepy site of Sunset Hills Farm. Reston is now the hub of the Dulles Corridor, and an alternative to Tysons Corner as a center of high-technology employment. It is no longer a quiet, uncongested town. It boasts one of the densest concentrations of high-rise office buildings in Virginia.[46] Reston's present vitality also depends on access to other places in the region for work, recreation, and employment—a modification of Simon's original vision.

THE BELTWAY, AGAIN, 1964

In 1964, when the last segment of the Beltway opened, the growth of Reston, Dulles Airport, and the Dulles Corridor lay in the future. Of more immediate interest were the properties around the major Beltway interchanges in Virginia. Traveling counterclockwise from the Potomac, these were, in order, Routes 123 and 7 at Tysons Corner, Route 50, Route 236 at Annandale, Braddock Road, and the Shirley Highway. There were other, smaller interchanges, but these were with major, established highways. Each interchange, when opened, immediately presented real estate developers with an opportunity to transform the landscape. President Eisenhower and others thought of the Interstate Highway System as one that would funnel traffic into and out of the central city. The Beltway's opening revealed a possibility—really, more of a certainty, and one foreseen by Frank Lloyd Wright—that a new form of development would occur at these interchanges. Development did occur at each, but among them Tysons Corner got the most. All five saw the development of housing, retail complexes, and corporate offices. To a lesser extent, all five attracted military contractors (which this study will be examining). But corporate offices grew to a significant concentration only at Tysons Corner. Tysons also became a magnet, attracting firms located either closer the Pentagon, or at other Beltway interchanges. That phenomenon sets the character of Tysons Corner more than any other factor. The following chapters will examine how this pattern emerged.

Among the Operations Research groups established after the end of World
War II was the Army's Operations Research Organization (ORO), whose
origins and Maryland location were described in Chapter 2. During the
Korean War ORO's stature increased, with employees conducting research
on the effectiveness of Army tactics deep in the field, sometimes encounter-
ing enemy fire. By the end of the 1950s, however, its relations with the Army
were strained. The issue concerned the degree to which the ORO involved
itself in the political, as opposed to purely mathematical, aspects of operations research. The Army wanted to replace its director, but it was rebuffed
by Johns Hopkins University, which administered the contract. The result
was the dissolution of ORO and its reincarnation as the Research Analysis Corporation (RAC), a Federally Funded Research and Development
Center.[1] Its new director, Frank A. Parker, talked to a developer he knew,
Gerald T. "Jerry" Halpin, about a location for the firm's headquarters. Halpin
suggested a site near Tysons Corner he was developing called Westgate
Research Park, across the Potomac from Parker's Maryland home, but easily
accessible from Maryland via the new Beltway bridge. Research Analysis
Corporation moved there in the early 1960s, one of the first of what would
become a flood of such firms settling in Tysons in the coming decades.

The story of RAC condenses much of what happened in Tysons Corner
in the early 1960s. This description sets the stage for our discussion of the
subsequent rapid growth of Tysons Corner.

In his classic study, *Edge City: Life on the New Frontier*, Joel Garreau mentions that the first defense-oriented firm to locate in the remote suburbs of
Fairfax County was not ORO, but Melpar. Named after its founders Thomas
Meloy and Joseph Parks, Melpar moved to a modern-looking building just
off Route 50. Surrounded by trees, greenery, and discreet parking lots, and

completed a decade before the opening of the Beltway, the building was an anomaly. Even today it looks modern, although with a distinctive, retro look. As of this writing, Melpar, now a division of Raytheon, still occupies the building. When the company located there, the movement of technology companies and Pentagon contractors to suburban Virginia was under way, but not as far out as Tysons. Many companies were moving to suburban offices closer in, mostly near Bailey's Crossroads and directly west of the Pentagon. Melpar was five miles beyond Bailey's Crossroads, finding itself just inside the Beltway when that road opened ten years later.[2]

Garreau and others have written about the architecture and setting of the Melpar building, but Garreau fails to note that Melpar's decision to locate in Falls Church did not lead to a concentration of similar companies nearby. Today the Route 50 interchange is home to several modern office parks, including one that hosts John "Til" Hazel's law office. The aerospace and defense contractor General Dynamics moved its headquarters there from St. Louis around 1990, but not many people work there, and few General Dynamics subsidiaries are located nearby. Likewise, in 1990 Mobil Oil, another Fortune 500 corporation, moved its headquarters from New York to an office park near that same Beltway interchange. But like General Dynamics, Mobil's presence did not have much impact on the character of the area. After merging with Exxon in 1999, ExxonMobil is now based in Texas.

To the south of this interchange, the Shirley Highway (I-95) interchange with the Beltway also offered potential for growth. After all, I-95 was the major north-south highway for the East Coast. In 1959 Atlantic Research Corporation (ARC), a company devoted to the design and manufacture of solid-fueled rockets, relocated from downtown Alexandria to a new, modern building in a wooded area near the future site of this intersection. Gerald Halpin, the developer of Westgate in Tysons Corner, was involved with this decision as well. Before becoming a Tysons Corner developer, he had been an Atlantic Research executive and was in part responsible for finding sites for that company's facilities. In the course of his work he had a chance to visit other research-oriented defense facilities, which he recognized were unique. The inherent value of these companies was not found in the machine tools, production lines, and raw materials that were the assets of the defense and military plants from the second World War. Their value was in the highly educated people who worked there. And every evening they left the building, presumably free to go elsewhere if they so chose. As one executive later put it, "Our assets wear shoes."[3] Halpin recognized the necessity, not luxury, of providing them a workspace that was pleasant and supportive of their creativity.

Atlantic Research, like Melpar, was such a company, founded in 1949 by an MIT-educated scientist named Arch Scurlock. Like Melpar, its main—almost only—customer was the Defense Department, and the percentage of employees with advanced degrees was high.[4] To attract and retain those employees, ARC's new headquarters was in a rustic setting, with windows facing the woods and an architecture that looked to the future. The building's design caused quite a stir, and even some resentment, among Alexandrians. In the 1950s Alexandria was a colonial Virginia town that claimed George Washington not only as one of its early residents, but also as one of its architects. One wing of the ARC building, visible from the highway, was roofed by a compound curve, similar to Eero Saarinen's design for the TWA terminal in New York's Idlewild Airport. Saarinen was not the architect for the ARC building, however, although he was asked.[5] The direct inspiration for the building may have been the headquarters for the System Development Corporation (SDC), in Santa Monica, California, whose *porte cochere* was a dramatic hyperbolic paraboloid, adopted as SDC's logo, and nicknamed the "flying diaper" by local residents.[6] System Development Corporation was itself an outgrowth of RAND. RAND's headquarters was not in the woods but was only a block from the beach, known as Muscle Beach because of the bodybuilders who lifted weights there. Executives from Atlantic Research had frequent contact with RAND and other southern California defense firms and must have noticed the architecture of their facilities. RAND's own corporate history specifically mentions the thought—and even mathematical theory—that went into the design of its Santa Monica headquarters, with the goal of creating "chance meetings" among its researchers in the hallways.[7]

In spite of this dramatic building, the Shirley Highway–Beltway interchange did not attract other companies of a similar caliber. Halpin left Atlantic Research, turned his attention to real estate development, and shifted his focus north to Tysons Corner. Around 1990 Atlantic Research relocated its headquarters to a modern but otherwise undistinguished building in Gainesville, Virginia, thirty miles farther west, where it had earlier secured a large tract of land for the dangerous processing of rocket fuels.[8]

Other aerospace companies have built branch offices at the Shirley Highway interchange, but on a modest scale.

Both the Shirley Highway and Route 50 interchanges pale in comparison to Tysons Corner. The same holds for the other Beltway interchanges: Route 236 (Little River Turnpike) and Route 620 (Braddock Road) in Annandale. Light industries, shopping, office parks for technology companies, government offices, and residences have emerged at these interchanges, but at a lower density. All the western Beltway interchanges became centers of commerce, but of them, only the Tysons Corner interchanges became an edge city.

In describing this phenomenon, Joel Garreau and the local press focus on the role of Til Hazel and his partner Milton Peterson, the developers responsible for the transformation of the county. Hazel's role was, and continues to be, important, not only for his personal role in developing farmland, but also for his tireless advocacy of what he feels is the best use of that land; namely, for residential, commercial, and corporate use. It has made him beloved by the development community, and he has forged close alliances with the heads of many of the corporations who have located at Tysons Corner. He has also made enemies among slow-growth advocates, as Garreau's book well documents. The irony of his prominence in Garreau's study is that although Hazel's work can be found throughout northern Virginia, it is concentrated along the Route 50 corridor: from the Beltway interchange where his office is located to the west through the center of old Fairfax City—now surrounded by suburban development—to the interchange with I-66, the site of the massive 'Fair Oaks' and 'Fair Lakes' complexes, and beyond. He has less of a presence in Tysons Corner, although he did develop the headquarters of the quintessential Tysons Corner firm Planning Research Corporation. Nor was he a major developer along the Dulles Toll Road through Reston. Hazel is nevertheless an important factor in the growth of Tysons Corner, as one who has provided housing, schools, and shopping for those who flocked to Tysons Corner for jobs.

The geographer Shelly Mastran wrote a detailed analysis of the development at each of the aforementioned interchanges, and her study reveals what are more direct factors.[9] She argues that no single factor can explain what happened. The development at Tysons and at the Dulles Corridor was the result of some dedicated actions by a few key individuals, plus events that can only be described as random, for which providing a purpose would be false to history.

Mastran examined courthouse and other land ownership records and determined that before the Beltway was completed, Tysons Corner had more available land for development (1,450 acres) than any of the other interchanges. It was mostly farmland, thus easier to develop for commercial, industrial, or residential uses than the other tracts, on which light industry or low-density housing had already been built. Of the available land at Tysons, most (512 acres) was held by a single family, the Ulfelders, who ran a dairy operation there beginning in 1925. A second large parcel, 187 acres, was held by Marcus J. Bles, who also farmed and had a large gravel pit at the crest of the hill. Besides those two landowners, four other families controlled large tracts, so that the top six landowners controlled two-thirds of the developable land at Tysons.[10]

What is fascinating about this ownership pattern is that it was just right. Tysons Corner had large tracts of land that developers could acquire for ambitious development schemes. But at the same time there was no single dominant landowner whose own wishes might hinder such plans. The landowners played a role in the conversion to commercial and residential use, but it was only through the efforts of dedicated and energetic developers and land speculators, who played a critical role in assembling parcels of land and holding them through the rezoning process, that something on the scale of Tysons Corner could happen. By themselves, the landowners lacked the resources. Such was the case at the Route 50 interchange, where nearly all the land, more than 1,300 acres at all four quadrants, was controlled by a single individual, Earl Chiles. Chiles was not a passive owner; he sold land to Melpar and for the Fairfax (now INOVA) Hospital. But in the early phase of land conversion, before the Beltway opened, he did not work with developers in planning for the systematic conversion of his holdings. According to Mastran, ". . . Chiles was generally not interested in selling his land, and when he did sell, was very selective about the new land use."[11] The remaining land was held in small parcels, few of which could easily be converted. The interchanges to the south had already seen commercial and industrial development before the Beltway came through. In hindsight, they had less flexibility for future growth than Tysons Corner had. Nor did these intersections attract, on the scale that Tysons did, land speculators who could buy these small parcels and assemble them into larger tracts.

Of all the interchanges, the complex at Tysons Corner had by far the most "open" land: dairy farms, gravel pits, little that stood in the way of future

FIGURE 4.1

Tysons Corner, circa 1965, looking southwest. The Beltway cuts through the picture from upper left to lower right, and the Dulles Access Road heads off to the upper right. Gravel pits and dairy farms still cover much of Tysons Corner at this time. Note the Dulles Access Road, foreground, stops at Route 123. The five-sided building in the center, just below the Beltway, was the home of the Research Analysis Corporation—the first Operations Research firm to move to Tysons Corner. Courtesy Fairfax County Library Photographic Archive.

development.[12] That statement assumes that open farmland is more suitable for development than land on which some commercial or residential development has already taken place. Therein lies a paradox: why not simply continue developing land that already has shifted away from agricultural use? One answer is that a developer can approach a dairy farm as a clean slate, with no need to undo what a previous builder has already done. The light industries and inexpensive housing that sprouted along the other interchanges were incompatible with the high-technology offices and upscale shopping centers planners saw in Tysons' future. Farms, on the other hand, were good only for farming. Virginia developers looked at the vast spaces available in the Midwest and said that farming was not appropriate for a place ten miles from the White House. That land's development in turn implies that the Fairfax Board of Supervisors was willing to rezone the land from agricultural to commercial or residential use. Such rezonings did occur. The many struggles attempting to prevent that from happening will be discussed later. (This does not imply that such struggles were insignificant. One can see them going on today in the outlying farming regions of Loudoun and Fauquier counties). Meanwhile, Tysons Corner has matured to a point where its future will depend on its existing buildings being torn down, at great expense, to make way for taller, even more ambitious buildings.

Gerald Halpin was among the first to recognize the advantages of the Tysons location. He moved his offices there from Springfield, Virginia and cofounded the Westgate Corporation in 1962. The name came from the perception that growth would occur to the west, toward Dulles Airport, which it did. Joining him was Colonel Rudolph Seely, Ulfelder's son-in-law, who was managing the family's dairy farming. The new Beltway almost exactly bisected the Ulfelders' dairy operations, which made working the farm in its final days an awkward activity.

Halpin and Seely were not alone. Marcus Bles had preceded them in speculating in the area. Bles bought his first parcel of land in 1949, and later added to his holdings by purchasing more land at Gantt Hill, a high elevation near the corner, where he established a gravel pit. In 1968 he sold about 100 acres of his holdings to a computer leasing company called Leasco. This company planned to use about half the land for its offices, and sell the rest at enough profit to effectively get its own holding for free. However, Leasco ran into financial difficulties. It was one of the leading examples of the "go-go" computer companies, whose run-up on the stock market and subsequent collapse anticipated the Internet bubble of the 1990s.[13] The collapse of the computer leasing business, combined with one of Fairfax County Board of Supervisors' periodic attempts to slow down growth in the county, forced them to sell their holdings by 1979—just when the market began to heat up again. Had they been able to wait a year or two to sell they might have doubled their return, but they were not so lucky. Leasco had bought the land from Bles for about $1.80 per square foot, and sold the last parcel for $6.50 per square foot. According to one of their executives, they soon thereafter started getting a steady stream of phone calls from people offering $12 per square foot.[14]

Another developer who bought land from Bles, Max S. Kraft, had a similar vision but even worse luck. In 1965 Bles sold Kraft and his partner fifty-eight acres, including the highest elevations of Gantt Hill, where the developer envisioned an upscale complex of apartments in high-rise buildings. Tenants would enjoy a spectacular view of the Blue Ridge to the west and the District to the east. The complex, called the Rotonda, eventually opened and has been the success its planners hoped for, but Kraft was forced out of ownership early in the process and never got the financial reward for his efforts.[15]

These examples chronicle how the land holdings of Bles and the Ulfelders were developed. But developers found ways to assemble the smaller parcels as well. A former executive with the Marriott Corporation, Frank C. Kimball, cofounded the Suburban Development Corporation and began negotiating with Tysons Corner landholders to acquire land for a regional shopping mall. While at Marriott, Kimball saw the success of Mariott's operation of one of its Hot Shoppes restaurants in Wheaton Plaza, a Maryland mall that opened in 1960.[16] There was already at least one suburban mall in Virginia, at Seven Corners inside the Beltway. But Kimball envisioned something more ambitious. His plans for a mall in Tysons met with resistance, and the Corporation had to contest with a number of other rivals. Some of the holders of small but critical parcels of land refused to sell; Kimball responded by allowing them to join the Suburban Development Corporation and participate in the profits it would generate.[17] The Corporation brought in the Wheaton Plaza builders—Gudelsky, Lerner, and Ammerman—to develop the Tysons Corner site, and in 1968 they opened the Tysons Corner Regional Shopping Center. With 1.5 million square feet of space, it was one of the largest in the region, and one of the first Virginia malls to have the major department stores that residents of the District and Montgomery County, Maryland had known for years.

Thus as Gerald Halpin saw better opportunities for science-based firms to the north of his Springfield, Virginia operations, Maryland developers saw better opportunities for shopping malls to their south. It was not just retail shopping that crossed the river into Virginia, however, as we saw from the Research Analysis Corporation's move to Tysons Corner. When this phenomenon began, the Maryland suburbs were more settled than their Virginia counterparts. Maryland and the adjacent upper northwest quadrangle of Washington had better neighborhoods catering to middle and upper class residents, better shopping centers, and most important, better public and private schools—the one factor real estate agents consistently cite as pivotal in families' decisions about where to live. Northern Virginia would eventually get all of those things, even surpassing Maryland with the improvement of its school system and the opening of a second Tysons shopping mall, but that came later.[18] Until that happened, and thanks to the opening of the Beltway's upper crossing of the Potomac in December 1962, executives and researchers could commute to office parks in Virginia, yet still live in Montgomery County, Maryland or northwest Washington. And the complex at Tysons Corner was the first major Beltway interchange they would encounter, less than four miles from the state line. Later on these factors would be less important, but they did play a role in starting the process, which, once started, was hard to reverse.

Combine the conditions above with the factor that has already been discussed: the 1958 decision to relocate the region's airport from Burke to Chantilly, with an access road to the Beltway at Tysons Corner. Earlier plans called for an elaborate system of highways to provide access to Burke, leaving the Beltway at the Braddock Road interchange, which was built to high standards in anticipation. The airport access road was to have been part of a network of new highways, including one to Thomas Jefferson's home at Monticello.[19] When the Burke site was abandoned, most of these roads were never built. With those highways scuttled, it was inevitable that developers' attention would shift elsewhere.

How much of this was planned and how much was serendipitous? In hindsight it seems obvious that things would turn out the way they did. Some of the people interviewed for Mastran's study told her that anyone could have made a lot of money by speculating on land in Tysons Corner at that time. But the examples of Leasco and Max Kraft show that those who saw too far into the future did not do well. Mastran notes that some developers lost money, and some even went bankrupt.[20] And Til Hazel's battles with local activists show that many residents of Fairfax County opposed nearly every attempt to develop the land, preferring to see it remain in dairy farming.

One factor not mentioned in these studies was the crucial role of the developers in cultivating clients with particular needs. Halpin recognized this when building the Atlantic Research Corporation's headquarters. The building had to fit in with the overall corporate goal of attracting and retaining good people. At Tysons Corner, other complicating factors came into play. The companies that located there were typically military contractors. They needed attractive facilities but at the same time did not want to look extravagant or call attention to the work that was done inside. The Pentagon set the pattern. It was a solid building and reflected the wealth and power of the United States military establishment, but at Roosevelt's insistence, it was not ostentatious. The buildings that housed military groups also were functional, attractive, and used quality materials, but did not look expensive. There would be no soaring glass atria or offbeat, postmodern architectural features.

Occasionally a building would violate these rules, with ill effects. Passengers driving along the Dulles Access Road cannot help but notice the headquarters of Virginia's Center for Innovative Technology (CIT) peeking over the treetops at the entrance to the airport.[21] CIT was partially funded by the Commonwealth as a state-charted, nonprofit corporation, and moved there in the mid-1980s. Its headquarters was designed after a nationwide competition among architects to fashion a structure symbolizing new directions in twentieth-century technology. The winning design succeeded too

FIGURE 4.2

"The Machine in the Garden": Headquarters of the Center for Innovative Technology, Herndon. For many years this was the first modern building a visitor saw when leaving Dulles airport. Photo by author.

well. The building can best be described as an upside-down glass pyramid, and it is hard to ignore. While Atlantic Research's hyperbolic paraboloid raised a few eyebrows in Alexandria, CIT's design suggested to Virginians that their tax money was being wasted. Never mind that the Center was a good idea, and has made a solid contribution to the local economy—the building just did not look right. Likewise, when in 1994 the secretive National Reconnaissance Office (NRO) moved to its new facilities in Chantilly, the press had a field day talking about the expense of the building, especially its bathroom fixtures.[22] The agency's very existence had been classified before 1992, and it was not used to dealing with public (or Congressional) scrutiny of how it spent money. As with the CIT, the NRO's problem was not that it had built a bad building, it was that it had failed to gauge the public's reaction to the design.

These two buildings are located far out on the Dulles corridor and were built after the initial development of the region. In Tysons Corner one or two buildings have attracted similar attention, but they are the exception rather than the rule, and were also built more recently. One is the Tycon Tower, one of the tallest in the region and prominently visible from the Beltway. It is a tall, brown, rectangular tower, with two loops at the top. This has led people to call it the "shopping bag building," which is all the more apt as it is located adjacent to the Tysons Corner Center Shopping Mall. But the building is attractive and otherwise modest, and whatever ribbing it gets from drivers along the Beltway is good natured. Likewise, the unusual loop at the entrance to the Tycon Courthouse building, at the corner of Route 123 and Old Courthouse Road, elicits chuckles, puzzlement, and a few jokes, but nothing scandalous.

In addition to navigating this minefield of public perception, builders had to meet their tenants' other needs. Because of the nature of their work, many tenants did not want anyone to know who they were or where they were located. That desire runs counter to a real estate developer's desire to advertise his or her success in renting a building. So while many Tysons buildings have the logos or initials of their major tenants on the top floors, many have no markings whatsoever. Investigative journalists uncovered who these tenants were, mainly by careful digging through real estate records in the Fairfax County courthouse, and for about ten years that information was available to the general public. But like the knowledge of the government's secret alternative sites in the Blue Ridge, this information is once again

FIGURE 4.3

CIA Headquarters, Langley, late 1960s. The view is to the north, with the heavily wooded Maryland suburbs in the background. Photo by author.

regarded as sensitive, and since September 11 it is no longer easy to obtain. It is impossible to hide the location of the Pentagon or the CIA headquarters, whose employees were attacked by terrorists on September 11, 2001, and January 5, 1995, respectively.[23] It has long been known, for example, that the CIA's Directorate of Science and Technology had facilities away from its headquarters. One published account mentions that the Directorate occupied a building in Rosslyn; other accounts mentions leases of several buildings in Tysons Corner.[24] In many cases the actual tenant is not the government agency itself, but rather a private, for-profit company who is contracted to do the intelligence work exclusively in one of those buildings. Thus a search of real estate records would not reveal the nature of the work being done there. There is no line, not even a fine one, separating the two.

Private contractors as well as government entities had practical reasons to keep costs low. They would each be competing with one another for government contracts. This naturally led the would-be contractors to have spartan facilities keeping costs down and allowing them to underbid their competition. They also did not want their facilities to seem out of line with the simple offices in the Pentagon, where most contracting officers worked. The successful developer learned how to juggle these incompatible demands.

As Tysons Corner and the Dulles Corridor evolved, the region became home to companies with fewer ties to government contracts. They were free to design expensive, ostentatious buildings that reminded passersby of their power and glory. They were proud to put their name on the outside of the building. One example is the headquarters of the Gannett Corporation, publisher of *USA Today*, which moved to a spectacular, "preposterously big," glass-faced building in Tysons Corner in late 2001.[25] Gannett is not a government or military contractor. Before 1990 these buildings were the exception to the rule.

Companies working for the government were required to include a place inside their buildings where classified information could be discussed. This was not done in an ad hoc fashion, but had to be carefully engineered and pass a series of security tests. These spots took on an acronym, which like so many is well known in northern Virginia but little known elsewhere: SCIF. It stands for "Sensitive Compartmentalized Information Facility" and is pronounced "skiff." The name implies that it is not simply a matter of information that is secret, top secret, or classified at even higher levels (whatever they may be). It is that certain information is to be divulged only to those in a specific "compartment," or specialty, and not to others, regardless of what security clearance they have. Initially the term referred to information gathered by intelligence agencies, including the CIA, NSA, and NRO. For these agencies the most sensitive information they possess is about what they call sources and methods, and these must be guarded above all others. Lately the term has taken on a more general definition. One may assume that the existence of a SCIF inside a building was itself a secret, as such knowledge suggests who the tenants might be. However, in 2003, a billboard off Interstate 66 in Fairfax advertised that the vacant building behind it was ready for lease and had a SCIF inside. Like the locations of the alternate seat of government, it is hard to keep some things secret in northern Virginia.

To qualify as a SCIF a number of rules must be followed regarding the type of doors used, who is allowed access, what kinds of telephone, video, and power cables are routed in and out and how they are secured, the existence and placement of ventilation ducts (a favorite access point in

Hollywood films), and so on. A SCIF would not have windows, although many are built behind false windows to preserve the typical windowed office façade and not give away what is inside. Special attention is directed to access: not only what kinds of keys, fingerprint readers, or other identification is needed to get into the room, but also how to prevent an unauthorized person from "piggybacking"—getting into the room by closely following a person with authorized access. The image of two people entering a room this way conjures up episodes of the Marx Brothers, who first perfected the technique, but it is a serious problem. One of the country's foremost SCIF providers is United American, Inc., which is located in Tysons Corner.[26]

The architect Christopher Alexander wrote an influential book, *A Pattern Language*, in which he describes a set of patterns, or general rules, that he believes are the key to good architecture. Tysons Corner developers came up with their own set of patterns, very much in the spirit of Christopher Alexander, although he might object to such a use of his name. These may be summarized as follows:

1. Pleasant surroundings, with lots of trees, water features, expansive lawns, or green space
2. Every scientist's or engineer's office provided with a window overlooking a natural setting
3. Lots of free parking, but visually blocked by berms or other structures to preserve the aesthetics of the setting
4. The use of high quality materials and solid construction, but in a way that does not call attention to itself or look too expensive
5. Good security, including restricted access to parking, and SCIFs when required
6. An extensive telecommunications and computer infrastructure. This amenity eventually included laying fiber optic cables under the local roads, giving the area possibly the most advanced telecommunications infrastructure in the world.[27]

With the rare exception of the occasional extravagant building, most of the developers in the Tysons Corner area knew how to meet their clients' needs. In the end they effected a transformation of commercial development not only for Tysons, but for the whole Washington D.C. region. These buildings have, in effect and without any conscious direction, become the region's signature form of architecture.

BAILEY'S CROSSROADS, 1960–1969

Not long after the Research Analysis Corporation moved to Tysons Corner, other research-oriented firms followed. Initially they came not to Tysons Corner, but to Bailey's Crossroads, the intersection of Columbia Pike and Route 7 just four miles to the west of the Pentagon. Offices of a private contractor, Computer Sciences Corporation, as well as three Federally Funded Centers—System Development Corporation, MITRE, and Analytical Services, Inc. (ANSER)—were located at Bailey's Crossroads in the early 1960s. Few of them remain there, but it is worth discussing them briefly as they have a continued presence in the region.

Computer Sciences Corporation (CSC) was founded in southern California in 1959 by employees of the United Aircraft Corporation, whose initial contract was to develop a computer programming language for business applications. That led to COBOL (Common Business Oriented Language), which, with the backing of the Pentagon, became the standard mainframe language for business and accounting into the 1980s. CSC branched into more general software support for engineering as well as business applications, with a concentration on aerospace and defense customers.[28] Their headquarters remain in southern California but they have a large presence in Tysons Corner and Reston, as well as in the Maryland suburbs. They no longer have a major facility at Bailey's Crossroads. In 2003 the company acquired DynCorp, a Reston company that was one of the original military contractors in the region, and one whose work went far beyond that of computer programming. (DynCorp's history will be discussed later.)

System Development Corporation's building in Santa Monica has already been mentioned as a possible inspiration for Atlantic Research's headquarters. The company was an offshoot of RAND founded to do research, development, and production of computer software for air defense systems—work that RAND preferred not to do as RAND felt that it would drain resources from its mission.[29] Beginning in October 1960, SDC established offices at Bailey's Crossroads and Alexandria, primarily for work on a system for the Defense Communications Agency.[30] The company grew and is widely considered to have founded the modern profession of computer programming. Alumni from its early military contracts went on to staff most computer programming jobs for years after. Its status as an FFRDC, however, caused problems for the company, especially when it entered into competitive bidding for jobs against for-profit companies. That led to a decision to convert SDC into a for-profit company, in 1969. Ten years later it was absorbed by the mainframe computer company Burroughs and lost its independence. For Burroughs the acquisition proved to be fortunate. By the mid-1980s the mainframe computer was being attacked in the marketplace by the nimble and far cheaper workstations based on the microprocessor. Companies like Burroughs, Sperry UNIVAC, and Honeywell could no longer compete; only IBM thrived as a mainframe supplier after several years of heavy financial losses. In 1986 Burroughs merged with Sperry to form the company Unisys, and it soon found that the systems work that SDC had pioneered was a good alternate source of revenue. Unisys's headquarters were in Detroit, where Burroughs had been, but by the late 1980s its northern Virginia offices probably generated as much or more business than any other divisions. Unisys is one of the top employers in the region today.

Analytical Services, Inc. (ANSER) was founded in 1958 under the sponsorship of the air force. Initially it had a staff of twenty five, doubling to fifty by 1961. All of its employees had security clearances.[31] Informally, ANSER was a smaller version of RAND: a Federally Funded Research and Development Center providing operations research and other mathematical analyses for the air force. Unlike RAND, it was located closer to its sponsor and therefore took on more mission-oriented, short-term tasks, which RAND avoided. ANSER stayed out of the public's view, again in contrast to RAND. After September 11 it took on the role of public spokesperson for the military's response to international terrorism, but it remains modest in size.

MITRE was founded in Massachusetts in 1958 to develop the nation's air defense systems.[32] That work evolved into an activity known in military circles by the acronym C^3I (pronounced "See-Cubed-Eye"), which stands for "Command, Control, Communications, Intelligence." Its roots go back to the British origins of OR, to manage radar installations prior to the outbreak of World War II. MITRE's work was oriented around the digital computer, especially the descendants of the famous "Whirlwind" computer developed at MIT in the 1950s. In the mid-1960s MITRE expanded its operations into other defense- and aviation-oriented activities that were computer-dependent, especially civilian air traffic control. By the mid-1970s it split its operations, with C^3I activities remaining in Bedford, Massachusetts, while other work moved to Alexandria and then to Bailey's Crossroads.[33] Around 1968 MITRE moved to a campus in the Westgate office park in Tysons Corner, which is in effect a coheadquarters for the company.

As with Systems Development Corporation, MITRE's status as a Federally Funded Center caused friction with the numerous for-profit firms that were settling nearby, who also competed for the same contracts in areas like air traffic control, and who had to pay taxes on their income. A 1978 Office of Management and Budget ruling was especially irritating: MITRE interpreted the ruling to mean that it was exempt from going through the competitive bidding process for work that was directly mentioned in its charter. This led to congressional investigations and to the establishment of a trade organization, the Professional Services Council, which argued that MITRE's status gave it an unfair advantage. The distinction is clear among these groups as to which firms are for-profit and which are federally funded, but to outsiders, and even to real estate developers, the distinction is not understood. A *Washington Post* article, for example, stated in 1978 that MITRE was the "largest private employer" in Fairfax County, with about 1,000 workers.[34] To the for-profit members of the Professional Services Council, MITRE was anything but private. It was a federal agency whose employees were exempt from civil service employment rules and salary schedules. In 1996 MITRE spun off a for-profit subsidiary, Mitretech, in response to this pressure.

Bailey's Crossroads was at the end of Columbia Pike, a highway that led directly to the Pentagon only four miles away. Although in the 1960s and 1970s it grew along with the rest of the county, Bailey's Crossroads did not become home to many more systems firms or defense contractors, as did Tysons Corner. Of the four firms mentioned that were there in the 1960s, only ANSER remains. Today it looks shabby and worn compared to the rest of the county.

The majority of companies that relocated to Tysons came from elsewhere in the country, especially from California. Table 4.1 lists some of the major for-profit companies founded in the west and relocated to Tysons or to the Dulles Corridor by the end of the 1970s. Most of them were founded in the mid- or late 1950s, in response to the tensions of the Cold War. In most cases, the companies' headquarters moved, but some, including SAIC, retained their headquarters in California and opened a coheadquarters in

TABLE 4.1
Western firms that relocated to or established a major branch in northern Virginia

Name	Where founded	Year founded	Year moved to Virginia	Subsequent developments	Main locations in 2002
BDM (Braddock, Dunn, and McDonald)	El Paso, TX (incorporated in New York)	1959	approx. 1964	Bought by Ford Aerospace, 1988; bought by Carlyle Group, 1990; went public, 1994; bought by TRW in 1997	Several Tysons Corner locations
CACI (California Analysis Center, Inc.)	Los Angeles Basin	1962	Early 1970s	A major contractor in Iraq war, 2003–	Arlington (Ballston)
CEA (renamed DynCorp)	Los Angeles Basin	1946	Late 1960s–early 1970s	Bought by CSC in 2003	Reston
CSC (Computer Sciences Corp.)	Los Angeles Basin (El Segundo)	1959	HQ remains in California; offices in Virginia from mid-1960s		Several locations in Maryland and Virginia
Melpar	New York	1940s	approx. 1945; also a branch in Cambridge, MA	Sold to various other companies after 1969; now a division of Raytheon	Route 50, Falls Church, since about 1954
Northrop Grumman	Los Angeles and New York	Northrop 1939; Grumman 1930; merged in 1994	HQ remains in California	Virginia presence after acquisition of descendants of PRC, BDM, and others	Numerous locations; largest private employer in northern Virginia as of 2005

PRC (Planning Research Corp.)	Los Angeles Basin	1954	1970 approx.	Bought by Emhart, Black & Decker, then Litton in 1995; Litton acquired by Northrop Grumman	Tysons Corner
R–W (Ramo-Woldridge); later TRW	Los Angeles Basin	1953	HQ remained in California	Merged with Thompson Products, 1958; bought by Northrop Grumman, 2002	Tysons Corner; Route 50 Corridor
SAIC (Science Applications International Corp.)	La Jolla, CA; later San Diego	1969	Washington, DC office in 1970; then to Virginia	HQ in San Diego but major branch in Tysons	Tysons Corner
SDC (System Development Corporation)	Santa Monica	1957	HQ remained in California, but opened Virginia offices in 1960	Acquired by Burroughs in 1980; now Unisys, HQ in Detroit	Divisions of Unisys in northern VA, including Tysons Corner

Virginia because of the volume of business conducted with the Pentagon. MITRE also followed that pattern, with its headquarters remaining in Bedford, Massachusetts but its Tysons offices handling an equivalent amount of work.

It is important to recognize the distinction between these offices and the common practice—by nearly every defense or aerospace company—of opening a storefront office near the Pentagon, where the company could show its flag, but did little substantive business. Little manufacturing of weapons hardware is done in northern Virginia. During the Cold War years that activity was concentrated in California, the Deep South, and New England, what geographer Ann Markusen calls America's "Gunbelt."[35] Yet because these manufacturing plants had essentially a single customer— Markusen calls it a "monopsony"—their owners felt a need to have a presence near the Pentagon, to keep open a convenient channel of communication and to get an early warning on new military doctrines that might lead to weapons development. The systems firms listed in table 4.1—including MITRE, CSC, and SAIC, who have their headquarters elsewhere—do not as a rule manufacture hardware. In fact, some, including MITRE and TRW, had clauses in their founding documents specifically excluding them from manufacturing hardware. Those clauses were there at the insistence of aerospace manufacturers, who did not wish to see competition established with federal government support. Their systems work involves integrating hardware that is produced elsewhere, and northern Virginia is primarily where this integration is done.

By the end of the 1960s the framework for the future growth of northern Virginia's economy was established. Defense contractors were located throughout Fairfax County, as well as in Alexandria, Falls Church, and Arlington, but Tysons Corner was attracting the most new tenants, and firms that had been in suburbs closer to the Pentagon moved out there. Dulles Airport had little traffic but did offer international flights, and with its dedicated access road leading from Tysons Corner it was convenient. With the opening of the Tysons Corner Center mall in 1968, residents could now find higher-quality stores than were available at other suburban locations, including the nearby malls at Seven Corners. Thanks to the efforts of developers like Til Hazel, professionals who worked at those systems companies had attractive places to live and good schools for their children. Fairfax County could compete with neighboring Montgomery County, Maryland as a desirable suburb. Homes with no indoor plumbing, which the Beltway surveyors encountered, were getting scarce.

And yet the region was still relatively empty. There were still dairy farms throughout the county. Reston was struggling to attract businesses. The Dulles Access Road saw little traffic. Developers of Tysons Corner properties still touted the area's low rents to attract tenants from the District and Maryland. What may be hardest for a modern observer to grasp was how uncongested and convenient the Beltway was. Traffic jams would come later, and by the time they came, in the 1980s, there would be no turning back. Tysons was by then the commercial center of the region and one of the strongest commercial and retail centers in the country.

FROM HERMES TO HERMES

Among the buildings one encounters while driving through Tysons Corner are those of 'Fairfax Square'—three low buildings on the west side of Leesburg Pike just outside the Beltway. The tenants on the upper floors include computer software firms and financial institutions, some of them replacing software and telecommunications firms that went out of business after the Internet collapse of 2000 and 2001. The tenants on the ground floor speak to the enormous wealth generated and spent in the region. One finds Tiffany, Gucci, and Hermes. That last shop, selling scarves favored by such customers as the Queen of England, is an elite among the elites. By some accounts, these stores on Fairfax Square generate as much revenue as the flagship shops in Manhattan or Beverly Hills.

Until recently, an office tower known simply as 8027 Leesburg Pike stood next door to Fairfax Square toward the Beltway. The more modern buildings nearby overshadowed it. In 2000 its tenants included a hodgepodge of trade associations, modest computer networking companies, real estate and law offices—and a place to buy pizza. That is typical of an office building near the end of its useful life, and indeed, in 2004 the building was torn down. Early aerial photographs show that it was one of the first multistory office buildings in Tysons Corner. Such buildings were rare, not only in Tysons Corner but throughout all of Fairfax County, before 1965. Its completion in the mid-1960s instantly made a lot of office space available in Tysons Corner. And the rents were cheap—less than $5.00 per square foot, according to several real estate developers—considerably less than what office space in the District cost. That price was a major factor in the movement of systems firms to Tysons Corner from elsewhere in the region.

Among those who moved into 8027 Leesburg Pike in the mid-1960s was Braddock-Dunn-McDonald (later known as BDM), a technical services firm founded in 1959 in New York. In 1960 it moved to El Paso, Texas, close to the U.S. Army's Air Defense Center at Fort Bliss, the White Sands Missile Range, and Holloman Air Force Base. At White Sands, the company provided technical advice to the army, which had established a missile firing range to rebuild and fire V-2 rockets captured from the Germans at the end of World War II. The Army had named that project after the Greek god who served as the speedy messenger to the other gods, Hermes. Most workers at White Sands pronounced Hermes as an American word: "*her*-meez."[1]

Joseph V. Braddock, Bernard Dunn, and Daniel F. McDonald each received a PhD from Fordham University in New York. At the time of BDM's founding, Dunn and McDonald were assistant professors of physics at Fordham, while Braddock taught physics at Iona College in nearby New Rochelle. In addition to teaching, each had experience either consulting for or as contractors with the Office of Naval Research, the Office of Ordnance Research, the Atomic Energy Commission, and the private firm EG&G. Collectively they fielded an impressive record of experience and expertise, especially in four major areas: analysis of missile systems, electronic instrumentation, radiation physics, and applied optics. A brochure they produced stated that "Feasibility studies, research and development, and program evaluation are conducted in the above fields. BDM's major activity is theoretical analyses in support of system test and development programs."[2] The three listed their experience and publication record in missile systems, guidance and control, instrumentation, and more specifically, the effects of nuclear radiation on missile systems. That focus remained remarkably constant for most of the company's history.

Shortly after its founding, BDM hired additional staff, growing to about a dozen professionals by 1962. Among those hired were Albert Lavagnino, a systems engineer with a masters from Harvard, and Earle Williams, an electrical engineer from Alabama with a degree from Auburn. Williams had met McDonald while working at the nearby Sandia Corporation, a nuclear weapons facility in Albuquerque. McDonald and his two cofounders recognized Williams's talent for management; by hiring him they could concentrate on the hands-on engineering they enjoyed more.[3] The combination worked: by 1968 Williams became BDM's vice president and general

FIGURE 5.1
Earle Williams, circa 1991. *Source*: Earle Williams

manager, and by 1972 he was president and CEO, a post he held for the next twenty years.[4]

BDM moved to El Paso because of the concentration of nearby missile-related activities. By the late 1960s Williams felt that although BDM was growing, there would never be enough business in El Paso to grow substantially. As the company was growing beyond the work provided by the nearby military installations, it attracted customers from elsewhere in the country. Williams felt that these customers did not think BDM could support them from its El Paso location. Not long after the completion of the Capital Beltway and Dulles Airport, BDM moved to 8027 Leesburg Pike.[5] Al Lavagnino may have been the one who selected that location; regardless of who decided, that building was one of the few office spaces of any size in Fairfax County at the time, and the decision to move to Virginia was a collective one. The founders do not recall whether they looked at Bailey's

Crossroads, where other defense contractors were located, but they did note Tysons Corner's low rents and easy automobile access to the Pentagon. Those two factors ruled out Montgomery County, Maryland. None of BDM's staff wanted to work in the District of Columbia. Williams, the Alabaman, made no secret of his preference for Virginia over Maryland. Maryland is south of the Mason-Dixon line, but Williams felt more comfortable in Virginia, even the part that had been occupied by the Yankees during the Civil War. The Fordham alumni who founded BDM did not object to Williams's preference, even if Fordham University lies in the shadow of Yankee Stadium in the Bronx.

So Tysons Corner it was. BDM, more than any other company, is most identified with the growth of Tysons Corner. From the late 1960s through the end of the century BDM and Tysons Corner grew in lockstep with each other. Not long after moving to 8027 Leesburg Pike, the company relocated to a larger space a few blocks away, on 1920 Aline Avenue (by coincidence, where a company called America Online would get its start a decade later). Then in 1977 it moved into its own high-rise tower on 7915 Jones Branch Drive, in the 'Westpark' office park developed by Gerald Halpin. By now BDM was growing rapidly, and tenants no longer moved to Tysons Corner for its cheap rents but for the concentration of scientific and technical talent located there. BDM built and occupied a twin tower next door only two years later, where it remained through the 1980s. Finally, still growing, it moved to a new facility on its own street, BDM Way, in the same office park. Five years later, in 1997, the company was acquired by TRW, and BDM lost its separate identity.

Throughout this period of rapid change some things remained constant. One was the nature of BDM's business. The percentage of work for the federal government remained around 90 percent, up until its acquisition by TRW. Of that 90 percent, most was for the Defense Department, although BDM had significant contracts with other federal agencies as well. And of those contracts with the Defense Department, many continued to focus on the Cold War problems that BDM worked on at White Sands: missile trajectory analysis and design, testing and evaluation of guided missile systems, the effects of nuclear radiation on missile guidance and electronics, design and evaluation of air and ballistic missile-defense systems, and simulation of war scenarios.

Like the Federally Funded Research and Development Centers it competed with, BDM built no hardware other than test equipment and prototypes. It could therefore claim that its evaluations of weapons systems were unbiased, as it did not have a stake in seeing that, for example, a Lockheed or Boeing airplane was part of the final weapons system.[6] With only one or two exceptions, that was also the case for the other companies that followed BDM to Tysons Corner. Little or no hardware is built in Tysons Corner, although software is another matter, and will be discussed later.

We encountered this hardware exclusion in the discussion of TRW, the company that bought BDM in 1997. TRW began as Ramo-Woldridge, founded in 1953 to do systems work for the air force on the Atlas missile. Because of objections from General Dynamics, which was building the Atlas, R–W agreed not to develop hardware of its own as it did the systems integration work. In practice, the company was able to evade or even ignore the ban, as it developed subsystems and electronic components for a variety of missile programs. A few years after its founding it received an infusion of funds from the Cleveland, Ohio auto parts manufacturer Thompson Products, who wanted to diversify into the aerospace field. In 1958 R–W merged with Thompson to become TRW, and with that it dropped the hardware exclusion entirely.[7] The federal government did not prohibit them from making that move. One reason may have been that General Dynamics and other missile manufacturers had all the defense business they could handle, and did not see TRW's competition as a serious threat. But there were objections among members of Congress and in the executive branch, leading to hearings and a detailed report on systems integration by the director of the Bureau of the Budget.

A major objection was that TRW used government funding and connections not only to leverage its way into the aerospace business, but also to make its founders wealthy when the company went public. According to some reports, the investment by Thompson was structured in such a way that it did not dilute the initial investment of about six or seven thousand dollars by Ramo and Woldridge. When the merger became effective in 1958, their stock was converted into shares of Thompson Products, and was then worth, on paper, around $3.1 million for each founder.[8] We saw how the Federally Funded Research and Development Center was founded to enable the government to pay scientists higher salaries than were provided by the civil service. The TRW experience raised this issue by several orders of magnitude: for those who founded such companies, wealth beyond any reasonable salary could come to them, if and when the company had its initial public offering (IPO) on Wall Street. The key issue here, which separates these companies from those in Silicon Valley, is the federal government's role as their almost-exclusive customer. If a company has business with the Pentagon that is all but guaranteed, one could argue that there is no element of competition—a factor dear to the theorists of venture capitalism. As one critic put it, the relationship gave the founders and directors of these companies the opportunity to have all the wealth of a capitalist enterprise, with none of the risk.[9] Of course there was competition among these contractors, and there was scrutiny from congressional staffers, but the nature of the business was different. The TRW issue would return in subsequent years to Tysons Corner, as many local firms followed its route and became public companies.

In Virginia, BDM continued the research-by-contract business model that it developed in El Paso. Its new location gave it the advantages of being located in a major metropolitan area, near an international airport and with access to a network of highways. Its proximity to the Pentagon enabled it to broaden its focus to the other service branches besides the army. Although the army was initially responsible for ballistic missile defense, as the capabilities of such missiles increased in the 1950s, the air force took over much of that work, with increasing budgets. Earle Williams described how the company grew from about $500,000 in sales when he joined, to about $4 million when they moved to Tysons Corner. Ten years after moving to Tysons Corner, BDM had ten times that revenue from contracts, and throughout the 1980s and into the 1990s, it grew by double-digit percentages nearly every year.[10] By 1992, the year Williams stepped down as CEO, BDM's revenue from contracts was over $400 million.

BDM's trajectory was set in motion by the Soviet Union's launch of an earth satellite, Sputnik, in October 1957. The Soviets, along with the Americans, had signaled an intention to orbit an earth satellite as part of the International Geophysical Year planned for 1957 through June 1958.[11] But few Americans had paid any attention. That a country perceived to be technologically backward could beat the Americans into space was politically unacceptable. The American response to Sputnik was multifaceted. Among those responses was the founding of NASA in 1958, as well as the

Advanced Research Projects Agency (ARPA), a defense agency tasked in general terms with ensuring that the country would no longer be surprised by such technological feats as Sputnik.[12] ARPA's role in creating the computer networking scheme that led to the present-day Internet has been told; its role fostering Internet-related work in the Dulles Corridor will be discussed later.

Less well-known were other steps taken by President Eisenhower in response to Sputnik. In addition to NASA and ARPA, the President's Science Advisory Committee (PSAC) was established, with James R. Killian (who was chairman of the Institute for Defense Analyses board at the time) selected as the first adviser. In August 1958 the President signed into law the Department of Defense Reorganization Act, which, in broad terms, shifted the balance of power within the Pentagon toward the Secretary of Defense and away from the individual services. Among the provisions of that act was the creation of a new post, the Director of Defense Research and Engineering (DDR&E), which further centralized oversight of research and development in the Pentagon.[13]

The 1958 restructuring of the Pentagon's authority came to the fore in 1961, when Eisenhower's successor, John F. Kennedy, appointed Robert S. McNamara as Secretary of Defense. During World War II McNamara had been with an operations research group that analyzed flight plans of heavy bombers to minimize fuel consumption as they delivered their ordnance to Axis targets. He was no stranger to the mathematical techniques discussed in chapter 2. When Kennedy appointed him, he was president of Ford Motor Company, where had applied those techniques to automobile production. McNamara seized the power implied by the 1958 reorganization and insisted on a top-down planning system for the acquisition of weapons systems. One of his first hires was Charles Hitch, an economist at RAND and president of the Operations Research Society of America. Hitch became the Department's comptroller—a position that had been obscure and little known, but assumed central importance under McNamara. With Hitch's help, McNamara devised a "Planning, Programming, and Budgeting System," which would apply mathematics and logical analysis to adjudicate among the services' individual desires to push through systems furthering their respective positions (which were without regard to the overall strength of the nation's defense). And McNamara insisted that he personally be involved in this process.[14]

This flow of power to the office of the Secretary of Defense further trans-formed the various Washington agencies and companies (and elsewhere) doing operations research and analytical studies. One would have expected that McNamara's initiative would benefit the Weapons Systems Advisory Group, the Institute for Defense Analyses, the Operations Research Office, and the other federally funded centers set up precisely for this kind of work. What happened was more like the familiar saying, "Be careful what you wish for; you might get it." WSEG and IDA got more work, but those in power did not have confidence in these agencies' ability to operate independently.[15] And the demands for detailed studies did not match their abilities. The indi-vidual services, faced with a need to justify their requests for new weapons, suddenly had to produce analytical studies of their own to counter those pro-duced by the Secretary's office.[16] A cynical view of this process was that the weight of the reports generated by each side had to balance each other before a program would be approved. And in spite of a few well-known cases of the Secretary cancelling a program, most of these weapons systems were approved and funded. In any event, the accession of McNamara resulted in a demand not just for more analytical studies, but for studies with narrow and specific purposes, tailored for specific audiences.

Robert McNamara served as Secretary of Defense until early 1968, when he resigned over the handling of the Vietnam War. The changes he brought to the Pentagon were too broad and too well entrenched to be reversed, either by his successor, Clark Clifford, or by President Richard M. Nixon's appointees. Among those changes was the slow decline in influence of the in-house analytical group, to the benefit of the for-profits.

What was distinctive about companies like BDM and others established in Tysons Corner around 1970? They were not the first to contract with the federal government on a for-profit basis. Daniel McDonald had worked for EG&G, a for-profit Massachusetts firm founded by the famous MIT profes-sor Harold Edgerton. Massachusetts was also the home of Bolt Beranek and Newman, founded in 1948 and later famous for its work in developing the Internet. TRW was likewise a for-profit firm whose systems division was located in California. What distinguished the Tysons companies was that they were for-profit, but at the same time they extended the work done by the nonprofit, federally funded operations research centers that grew up a decade earlier. Their origins go back to the Research and Development Board, which Vannevar Bush hoped would provide unbiased analysis to the military.

The RDB's shortcomings led to the Weapons Systems Evaluation Group, then to the Institute for Defense Analyses, the Operations Research Office, and other groups based in or near the Pentagon.

The Federally Funded Research and Development Centers were established to overcome the shortcomings of civil service hiring and salary restrictions. But they were only partially successful. The Tysons Corner firms injected the profit motive into this equation. Thus, as the Institute for Defense Analyses was established because of the perceived inadequacies of the Weapons Systems Evaluation Group, so did the for-profit companies establish themselves to address what they, and their Pentagon contracting officers, felt were weaknesses among the federally funded centers. Joseph Braddock has described how one of BDM's very first contracts was to test a security system already evaluated by IDA—to check and see if IDA was doing its job.[17]

Even before the issue of selling shares to the public arose, the profits made by these companies led to criticism. Sometime in the 1970s, critics began using the phrase applied to the thieves who used the newly opened Beltway to rob suburban houses. That phrase may have gained popularity not from the Washington papers but from a *New York Times* column in 1976, where these companies are mistakenly identified as "consultants." Criticism also came from within the government, including from Pentagon officers who felt that giving a contract to a for-profit entity implied that they lacked the competence to do it themselves.[18] Criticism reached a zenith during the Carter administration (1977–1981). That was ironic: President Jimmy Carter had served as an officer on a nuclear-powered Navy submarine, and he must have known the value of systems engineering. Indeed, from the time of their establishment in the late 1960s, these firms have been criticized as much by their peers in government, who want that work for themselves, as they have been criticized by those who object to the militarization of the U.S., economy. In any event, President Carter's administration is not remembered fondly in northern Virginia.

Joe Braddock and Earle Williams saw this criticism in a different light. At one time they found the Beltway epithet amusing, but they no longer do. They both have argued strenuously that whatever one called BDM's employees, they most definitely were not consultants. In Williams's words, "We never refer to ourselves at BDM as 'consultants.' Never."[19] Braddock argued that a consultant is one who analyzes a problem, writes a report suggesting a solution, submits a bill, and then walks away. BDM, in his words, stayed with the client until the problem was solved.[20] He likened the company to a law firm, which works with a client to solve a particular problem, for which it is paid. Of course lawyers, like consultants, are often the target of unflattering comments, but most, in Washington at least, acknowledge their value. BDM's emphasis on contracting also meant that there was no permanent bureaucracy created; once a job was done, the team disbanded and members were free to bid on other jobs. (Note, however, that Tysons Corner is home to at least one large firm that proudly calls its employees consultants: Booz-Allen & Hamilton, which moved there from Maryland in 1992.)[21]

In their annual reports and interviews with the press, BDM and its neighbors acknowledged the difficulty of describing what they did. That some of the work was classified added to the problem, but that was not the real reason. It was more the intangible nature of operations research and its descendants: systems analysis, systems integration, and the like.

BDM officers argued that, as a for-profit company, they lowered costs for the Pentagon. Like any for-profit company they sought to keep costs low and to work efficiently. They also refuted the charge that because they had only one captive customer, there was no competition. There could be plenty of competitors nearby also were bidding on the same contracts, although the need for specialized information regarding the arcane details of a contract often meant that other companies would not be able to make an effective proposal. Like the early contract to check on IDA, BDM often got contracts to evaluate the quality of work supplied by major aerospace or defense firms to the Pentagon. In these cases the aerospace giants were the villains, billing the government $435 for a hammer or hundreds of dollars for a coffee pot. When hired to watch for such abuses, BDM's competence saved the taxpayer money, Braddock argued.[22]

Another selling point, though I have not uncovered direct evidence that it was used, was that the for-profit companies could be relied on to be more discreet. Leaking sensitive information to selected journalists or congressional staffers is a way of life among federal employees. But if a private firm did that, it would immediately suffer financial consequences. As publicly traded companies, these firms had to disclose their financial data to the public, but they did not have to answer to the congressional inquiries that always dogged the federally funded centers. Those centers suffered during the Vietnam War era, especially after Daniel Ellsberg, a RAND employee, leaked a RAND report—the so-called Pentagon Papers—to the *New York Times*. Likewise, in the late 1960s a branch of IDA was driven from the Princeton campus by student protests, and in the 1970s the CIA had to endure embarrassing scrutiny over its alleged spying on domestic antiwar protestors. The for-profit firms would later on be exposed to scrutiny, but there have been few breaches of security or leaks to journalists coming from them.

The one place where these analogies with other professional services groups fail is the one already mentioned: the Tysons Corner firms sold their services mostly to the Pentagon. And not just "the Pentagon" as a synecdoche for the defense establishment, but "the Pentagon" as in the building where contracting decisions are made. In economist's terms, the situation is a "monopsony": many vendors, but only one customer. From time to time the companies attempted to sell their services to other customers, but many of those attempts failed, and they usually refocused on defense contracts.[23] That, more than any other factor, explains the concentration of these for-profit firms in Tysons Corner.

Among the many effects of a monopsony is that it negates the major advantage of contracting mentioned earlier: that after the work is done, the team no longer needs to be on the government payroll. These companies felt a need to remain close to their clients. Employees of NASA, for example, have privately remarked to me how these contractors establish a permanent presence at NASA centers, and it is impractical to send them away after a job is done. NASA is not a defense agency, but this applies to the armed services as well.[24] On many military bases, only the color of a person's badge indicates whether he or she is a contractor or government employee; there is no inherent difference in the work that each member of a team does. But there can be a dramatic difference in pay. The practice has its own acronym, 'SETA.' Well-known to military contractors but a mystery to outsiders, it stands for "Systems Engineering and Technical Assistance," and describes designating a contractor to "live" inside a federal agency and monitor its day-to-day work. These people provide the technical expertise that the federal agency can no longer afford to have. This is especially true of military units with frequent personnel turnover. Military officers are often sent to a new billet every three years or so, regardless of how much expertise they have built up. The best way to advance military careers is to be deployed overseas in combat or high-risk theatres, with only an occasional tour of duty in Washington. The use of SETA workers reinforces the perception that the government cannot attract people with the required skills in technical matters.

Many government employees take issue with this perception, but it will not go away. They use another, more pejorative, term for SETA: "body shops." This refers explicitly to the practice of bringing in contract workers who are de facto permanent employees, but because they are contractors they do not show up as full time equivalents (FTEs) on the federal payroll. In the mid-1980s, as this practice was at its zenith, the Washington business press criticized it, not for circumventing personnel rules, but for leading to a risk-averse climate in the region, especially when compared to Silicon Valley. The press argued that although Tysons Corner firms had as much technical talent as Silicon Valley, the Virginia firms would never take the kinds of entrepreneurial risks that led to famous start-ups like Apple or Intel—companies that profoundly changed the way we live and work. In Tysons Corner, critics argued, it was too easy to cultivate relationships with military contractors and generate steady, increasing, but unspectacular profits.[25] As Silicon Valley companies began to dominate the news and enter the public's consciousness, and after the Cold War ended in the early 1990s, this criticism took on a sense of urgency, implying that the northern Virginia economy was fragile and could be devastated by changes in military procurement. We shall see that the problem went away, but it remains a structural element of the local economy.

BDM and its brethren learned how to navigate the contracting maze. That required knowing the details of how contracting offices in the Pentagon worked, information obtained by hiring retired military officers. It also meant knowing how to get the required signatures on the proper forms, meeting obscure deadlines for filing, and paying attention to other details published in the fine print of the requests for proposals. The barriers for entry into this world were high. Once a company got over those barriers it made an effort to stay there. Nevertheless, capital requirements were low and the volume of work coming out of the Pentagon was increasing exponentially. In theory, the rules of the game were not secret and were equally accessible to all. Some joked that an individual needed two things to start a firm: a personal computer in his basement, and a pack of business cards. The truth was somewhere in between. This study concentrates on BDM and the other major systems firms, but the Professional Services Council, a trade association of these companies, listed over 120 members in its directories in the early 1990s. In addition to the large firms, many others were quite small, with only a handful of employees. Individuals operating out of their homes were not uncommon, though most members of the Professional Services Council were larger.[26]

The difficulty of advertising for a contract, receiving and evaluating bids, and making an award (one which the losers would not challenge) caused frustration at the other end as well. Pentagon officers found it difficult to quickly award a contract and get an urgent job done. That led to a practice whereby contracting officers made extensions to an existing contract to continue employing a firm, regardless of whether the new work was related to what the original contract was for. This issue boiled over in 2004, when CACI, operating on an extension of an innocuous Interior Department contract, was hired to interrogate prisoners in Iraq.[27] Another practice was to let a contract to a company owned by a qualified minority, under a so-called Section 8 (a) exemption that favored such companies in the competitive bidding process. The term came from the Small Business Act of 1958, which permitted the SBA to authorize a minority-owned business to enter into prime contracts with the federal government directly, bypassing the competitive bidding process.[28] None of the Tysons Corner firms discussed in this chapter qualified as Section 8 (a) minority-owned companies, but many of them did work as subcontractors to such companies. It was not hard to find charges, usually from those who lost a bid, that the Section 8 (a) company was a sham. However, if technical work had to be done, those who had the requisite skills got the job. The frequent criticism of these and other alleged abuses often failed to note that fact. Companies like BDM were staffed by technically competent people. And if an employee could not do the work, he or she could be fired—another way the for-profits differed from military and other government agencies.

LEGAL RESPONSES FROM THE WHITE HOUSE AND CONGRESS, 1962–1983

The rise of for-profit companies doing what had been done by government entities did not go unnoticed. One of the first official responses to the phenomenon was a report issued by the White House in 1962 titled "Report on Government Contracting for Research and Development," usually called the "Bell Report" after David E. Bell, director of the Bureau of the Budget and one of its authors.[29] That report addressed the issues in broad terms but did not focus on the companies like BDM, which were only a few years old at the time. Thus the Bell Report concentrated mainly on the large aerospace companies, but the issues it addressed would be relevant to the systems firms described in this chapter. For this study, the most important issue that Bell addressed was the loss of technically qualified people from the Pentagon and other federal contracting agencies, who thus lost the ability to judge the quality of proposals submitted to them. This issue remains relevant, and is one of the principal objections to the extensive use of contractors in the current war in Iraq. Another topic the report covered was military aircraft, or other weapons plants, that were owned by the government but operated by a defense company like Lockheed or General Dynamics. These Government-Owned, Contractor-Operated (GOCO) facilities were proliferating rapidly, and further represented an erosion of the so-called arsenal system of weapons production that had existed earlier in the century. None of the GOCO plants, as far as I can tell, were located in Tysons Corner or the Dulles Corridor.

The Bell Report was followed a few years later by a modest, eight-page typed document, known as Circular A-76, published by the Bureau of the Budget under the Executive Office of the President in March 1966. This document spelled out under what circumstances activities should be performed in-house by government agencies, and when they should be contracted out. It stated that "No executive agency will initiate a new start or continue the operation of an existing 'Government commercial or industrial activity' except as specifically required by law or as provided in this circular."[30] The threshold for triggering the application of the "new start" provision of this circular was "an activity involving additional capital investment of $25,000 or more or an additional annual costs of production of $50,000 or more." The Circular was updated in 1967, 1979, 1983, and 1999, at which times it was modified, extended, redefined, and debated. It remains in force.[31] It is no exaggeration to say that in the Pentagon and in other federal agencies, one can find offices whose inhabitants do nothing all day except parse the meaning of this circular.[32]

Federal agencies usually signaled an intention to let a contract by publishing a notice in the *Commerce Business Daily*. They did not have to publish the contract requirements, only the intent to let. CBD was published by the U.S. Department of Commerce and printed by the U.S. Government Printing Office every business day. It moved to electronic-only publication in the 1990s and is now called *Federal Business Opportunities*. The current notice states that it is "the single point of universal electronic public access on the Internet for government-wide federal procurement opportunities that exceed $25,000."[33] In theory this ensures fairness among all those wishing to do business with the government. There are conflicting stories about how this system actually worked. Employees at Melpar told me that they read the *Commerce Business Daily* every morning and would often prepare to bid on a contract based on what it listed. On the other hand, studies of the Advanced Research Projects Agency's role in creating the ARPANET, the predecessor to today's Internet, often state with pride how ARPA was able to let contracts on a very informal basis, based on personal relationships between university researchers and ARPA directors. Note, however, that the seminal ARPANET contract to Bolt Beranek and Newman, for the network's switching devices, was awarded only after competitive bids were solicited.[34]

Periodically Congress sought to keep the process open. After 1983, Congress required that agencies wait at least thirty days after publishing a notice in the *Commerce Business Daily* before negotiating with a contractor, to further ensure fairness.[35] The next year Congress passed the Competition in Contracting Act, which reaffirmed the requirement that contracts be awarded only after a competitive process. Some ex-ARPA officials recall that this act ended ARPA's days as an innovative funding agency, although other factors were probably at work.[36] Also after 1984, a regulation known as the Federal Acquisition Regulation (FAR) was applied across the board to all federal contracting, civilian and military. However, the Pentagon was able to retain some differences in this method of contracting, which vitiated some of this reform. In the 1990s there was an attempt to clamp down on abuses of the Section 8 (a) exemption. As with many such rules, in addition to the desired effect of keeping the process open, these rules had the collateral effect of generating new and creative ways of getting around them, thus raising the barrier to newcomers wishing to enter the contracting world.

The popular press often speaks of a revolving door between the private sector and the military. It is more accurate to say that there is no door at all. Although executives of these companies took offense at the Beltway epithet, they liked to joke about their competitors being burdened with retired military people. Executives told me, for example, that PRC stood for "Place for Retired Colonels," CACI for "Captains and Colonels, Inc.," and so on. Earle Williams disputed the notion that BDM brought retired military people in solely for their knowledge of the corridors of the Pentagon; he says that BDM sought them out because of the technical knowledge they gained while working with complex military weapons systems. He estimated that at its peak, about 20 percent of BDM's employees were retired military officers, while noting that the public perception was probably closer to 80 percent.[37] Throughout the Cold War era the military allowed someone to retire at 50 percent of their base pay after twenty years of active duty.[38] For those who graduated from the service academies or who were in ROTC in college, that meant retirement, if they chose it, at around age 45. These people were not typical retirees: they were in good health, had plenty of energy, and were mentally sharp. If, prior to retirement, they had served a tour of duty in the Pentagon, they might also have a house in the Virginia suburbs—with a corresponding mortgage—and children approaching college age. In short, they both wanted and needed a job and were willing to work hard. Their years in the military gave them a good work ethic: they were punctual, courteous, and dressed conservatively. They also tended to be obsessed with rank and title, a trait that had to be unlearned at a private firm that prized creativity and entrepreneurship. Federal rules against double-dipping would reduce their retirement if they took another government job, but that did not apply to the private sector. This was the talent pool that BDM and its counterparts could draw from.

Because of its rarity, economists have devoted less attention to the workings of a monopsony than they have to the more common monopoly or oligopoly. It is therefore difficult to judge whether the U.S. military receives the best possible service from these private contractors. The mission of the Defense Department is not to make or sell a product but to defend the United States, whatever the cost. Also remember that the Pentagon may be a single building, but the Defense Department is not monolithic. Vannevar Bush hoped that the Research and Development Board could adjudicate the rivalries among the army, navy, and air force. It failed to do so. The for-profit contractors face the same issues of fiefdoms and lack of coordination among the various defense agencies.

We may now return to the issue of stock ownership. With one or two major exceptions, the private companies sought to cash in on their initial investment by offering shares to the public, typically after a period in which they proved their economic viability and showed a string of profitable years. This IPO usually resulted in a financial windfall for the founders and any others who received large blocks of shares when the company was starting out. For computer and telecommunications companies, the IPO is much discussed, and it is considered one of the driving engines of Silicon Valley and the Internet economy. The IPO worked in the same fashion for government services firms.[39]

The process also had a downside: once a company went public, it was subject to the machinations of Wall Street speculators, of whom there was no shortage, especially in the 1980s. BDM felt both of these effects. Its founders became wealthy after it became a public company. Then, in 1988, it was acquired by Ford Aerospace in an attempt by Ford to move into the lucrative contracting business. Ford Aerospace had a long aerospace tradition going back to the famous Tri-Motor airplane, and it was a premier maker of communications satellites, but the fit with BDM was poor. Two years later, with the help of the Washington financial firm The Carlyle Group, the company separated itself from Ford and became independent again.[40] The Carlyle Group bought BDM from Ford for about one-third of what Ford had paid a few years before. When BDM was acquired by TRW in 1997, the initial investment was repaid more than sixfold. But even that transaction was not the end of the story: TRW itself was acquired late in 2002 by Northrop Grumman. These activities, though profitable for some of the players, were also unsettling for many of BDM's employees and customers. In many cases such mergers usually are followed by layoffs, but that appears not to have happened here. The net result was that BDM's workforce and its skills have spread through the region, as an invisible but nonetheless real foundation of the local economy.

The preceding discussion focused on BDM because of its role as a defense contractor and for Earle Williams's role in the economic development of Tysons Corner specifically. The details of the economic development will be discussed in the next chapter, but before turning to that topic it is worth a look at other major contractors who established themselves in Tysons Corner or nearby. The Professional Services Council listed over 100, among which a few stood out. These were California Analysis Center, Incorporated (CACI), Planning Research Corporation (PRC), Science Applications International Corporation (SAIC), and DynCorp.

CACI

CACI and PRC both have a direct connection to the RAND Corporation. In a previous chapter we discussed how CACI was founded by RAND employees seeking to commercialize simulation software they had developed. One of its founders, Harry M. Markowitz, had been hired by RAND to work on economic modeling that took the form of large matrices of numerical values. Around 1959 Markowitz moved to General Electric in New York, where he began developing simulation software for that company's manufacturing processes. He left GE and returned to RAND after a short stay, in part because he felt that although GE would use his work, it would not spread beyond the company. At the urging of Herb Karr, in July 1962 Markowitz, Karr, and Bernard Hauser founded CACI with the intent of selling their simulation software, SIMSCRIPT, commercially.[41] Gradually they shifted more and more of their time to the new company. Thus, the first version of SIMSCRIPT appeared as a RAND publication, but version 1.5 was sold as a proprietary CACI product.[42]

Markowitz later had a falling out with Karr over the business model for the company, and in March 1968 the board fired him. In 1972 Karr hired J. P. "Jack" London, an Annapolis graduate who brought navy-style discipline and a degree of professional management that the academics who founded the company needed. Also around that time the company moved to northern Virginia and began to concentrate on federal contracting. It is presently located in the Ballston area of Arlington, inside the Beltway about five miles from Tysons Corner. London rose to become president and CEO in 1984, and was for CACI what Earle Williams was for BDM: a capable manager who steered the company through cycles of ups and downs in federal contracting to a place where CACI is a major federal contractor. As of this writing CACI is one of the few of the original northern Virginia companies to remain independent.[43]

PRC

Planning Research Corporation (PRC) also had its roots at RAND, where a group by that name was established in the 1950s. It was founded in 1954 by three RAND employees: a physicist, an economist, and an engineer. Like CACI, it moved to the Washington area in the mid-1960s to be closer to the seat of government, but also because its founders had developed a substantial business in Europe and wanted better access across the Atlantic.[44] Like BDM, it built no hardware of its own. Also like the others, it began with skills in operations research, but that soon evolved into what is called "systems engineering." This term appears often in the discussion of federal contractors, and it lacks a clear definition. An informal definition, from a practical standpoint, is that systems engineering extended the techniques of OR into a general suite of software products that ran on digital computers. Systems engineering also has a more formal definition, that deals with the development of large-scale technological programs, such as an intercontinental ballistic missile (ICBM), in which the many components, produced my many subcontractors, have to mesh with each other to function as a whole. Like operations research, its origins also go back to World War II, to the development of fire-control devices for anti-aircraft weapons. Modern systems engineering is said to have originated with the Atlas ICBM project, for which General Dynamics was the prime contractor, but Ramo-Wooldridge was the firm established expressly to integrate the myriad and complex subsystems.[45]

In recent years these techniques have spread into every corner of society, and the term has been replaced by the vague, catch-all phrase "information technology," or IT. Beginning in the 1990s, IT became one of the most popular undergraduate majors on college campuses. It is less rigorous than computer science or computer engineering, and in the simplest cases involves little more than unpacking a shipping box, plugging in the power cord and networking cables, and loading appropriate commercial software. In practice, setting up even a simple home system is fraught with complexities that baffle the layperson. And installing such systems on a military platform, like a reconnaissance aircraft or submarine, is several orders of magnitude more difficult, with life-threatening consequences if not done correctly. All of this activity may be said to be the legacy of R-W on the West Coast, and its Northern Virginia counterparts.

So the products of a company like PRC were not just analytical reports and analyses but also customized software that ran on the customer's computers. Through the 1960s, that meant large, so-called "mainframes," supplied mainly by IBM, Burroughs, UNIVAC, and Honeywell. PRC and its counterparts could not afford a mainframe of their own, but they were able to rent time on a mainframe supplied by a service bureau or other time-sharing company, of which there were many in the late 1960s. At that same time, the mainframe was supplemented by smaller minicomputers, supplied by the Digital Equipment Corporation, Data General, and a few other vendors. These machines did not require special, dedicated, climate-controlled rooms, and the systems engineering firms could afford to buy their own on which to

develop software. Like the mainframes, the minicomputers also required programming—in some cases, they were more difficult to program because of
their more limited memory capacity or processing power.[46]

Around the time of PRC's move from California in the late 1960s, it
acquired a southern California civil engineering company, one of whose
employees was John Toups, a graduate of the University of California at
Berkeley. Toups moved with the company and rose to become CEO in 1978,
a post he held until 1987. Under his leadership, the company prospered, and
he, along with Jack London and Earle Williams, spoke for the northern
Virginia federal contractors during the peak years of their independence.[47]
Toups recalled that PRC did a lot of work for the Pentagon, as did the other
companies, but it also did a significant amount of contracting for other federal and local government agencies, and perhaps one-third of its business
was for commercial customers. One of its biggest military contracts was for
WWMCCS—the World Wide Military Command and Control System
(pronounced "Wimmicks")—a military communications network implemented mainly with Honeywell computers. A major civilian project was the
nationwide implementation of a 911 emergency phone number—a huge
and socially significant civilian contract, and one that PRC implemented
using computers supplied by the Digital Equipment Corporation. Another
nonmilitary contract was in the real estate profession, for which PRC implemented the Multiple Listing Service (MLS), still used by realtors to gain
access to up-to-date information on houses for sale.

Initially PRC moved to offices in the District of Columbia, but, perhaps in
response to the riots following the assassination of Dr. Martin Luther King in
1968, it moved to Tysons Corner. As with BDM, the developer and former
Atlantic Research executive Gerald Halpin provided a building, at 7670 Old
Springhouse Road in the Westgate development, just east of the Beltway-
Route 123 interchange.[48] A few years later PRC moved to a larger campus,
on its own PRC Drive, off Lewinsville Road toward McLean. This development was one of the few in Tysons Corner that was built by Til Hazel and his
partner Milton Peterson.

Around 2000, PRC moved away from that campus and back to the Westgate office park, to a building on Colshire Drive. By that time it had lost
its independence, although it was for a while listed as a named subsidiary of
Litton, the defense contractor that bought it around 1995. Before the Litton
acquisition the company went through a convoluted period in which it was

bought in 1986 by the Connecticut tool maker Emhart, which in turn was acquired by the Maryland firm Black & Decker, a maker of consumer tools, then spun off and sold to Litton. The merger with the tool makers was not a good fit, but the merger with Litton was better, and the combination proved successful. Litton itself was acquired by Northrop Grumman as part of the wave of post–Cold War mergers around 2000.

In the discussion of BDM's acquisition by TRW, the issue of these companies not making hardware arose. The aerospace giants were developing their own in-house systems capability, and a company that made no hardware was at a disadvantage in bidding on complex contracts that the giants could offer as a single package. PRC's loss of independence was related to that factor, though it also seems to have been a victim of Wall Street speculation. Once these companies go public, their shares can be acquired for all kinds of reasons, many unrelated to the company's core business. In PRC's case it may have been a desire by Emhart to diversify or to get a share of the federal contracting, even if the cultural fit was poor. The Black and Decker acquisition was likewise questionable and may have been simply to get rid of a competitor. Note that SAIC, another Tysons firm, has managed to remain independent, and it is not (yet) publicly traded.

SAIC

SAIC is not, strictly speaking, a Tysons Corner firm. The company was founded in San Diego and its headquarters remains there. However, it established offices in northern Virginia only a few years after its founding, and its physical presence in northern Virginia is as great as any other. Its Tysons Corner facilities serve as its East Coast headquarters and direct as much of SAIC's policy as its San Diego Facilities do. In 2004 it employed around 15,000 highly skilled and well-paid people in the greater Washington region, far more than in San Diego. Of that number, over 3,000 work in Tysons Corner.

SAIC grew out of contracts given to General Atomics for the construction and management of facilities for the national atomic energy and weapons laboratories. One of General Atomics' employees, a University of Michigan graduate named J. Robert Beyster, had been at Los Alamos and worked with the Nobel Laureate Hans Bethe. In 1969, after General Atomics was sold to Gulf Oil, Beyster left and founded Science Applications International

Corporation, operating out of an inexpensive office (albeit with a view of the Pacific) in nearby La Jolla.[49]

From the start Beyster instituted two policies that would set SAIC apart. One was that it would be employee-owned. Employees were awarded stock or would have the opportunity to purchase stock in the company. They could also sell shares back to SAIC, but they could not buy or sell shares on the open market. If they retired, they could sell their holdings back to SAIC or to other employees at a price set internally by the SAIC board, not by Wall Street. The company thus never experienced the flush of cash that often came with an initial public offering, but employee ownership also meant that it had a freedom and independence that its competitors did not. That freedom allowed it to buy (and sell) other companies, yet not worry about a big conglomerate or aerospace firm buying it, as was the fate of PRC, BDM, and others.

The second policy was that SAIC operated more like a holding company than as a single, unified research firm. This allowed its employees to seek out businesses that could take advantage of their technical know-how, but that otherwise had little to do with SAIC's original specialty in nuclear physics. Its recent annual reports do not even list nuclear physics as one of the company's "Key Vertical Markets."[50] In 2003 Beyster stepped down as CEO, and his successor, Kenneth C. Dahlberg, hinted that he might change that policy.[51] But these policies thus far account for the diversity of SAIC's activity and for its numerous offices all over the region: in Arlington, Alexandria, and along the Dulles Corridor, where its subsidiaries occupy at least ten separate offices as well as several locations in Maryland.[52] Its Virginia headquarters have been in Tysons Corner, in a large building on SAIC Drive facing Route 7, with a McLean postal address. It also occupies significant office space on Science Applications Court off Gallows Road on the southern edge of Tysons, with a Vienna postal address. This address is listed on older maps as Boeing Court, from a time when those buildings housed the aerospace company's computer services division.

Among SAIC's most famous acquisitions was the purchase, in 1995 and 1996, of Network Solutions, Inc., the keeper of the registry of .com and .org Internet addresses. That was at the beginning of the explosion of the commercial Internet, when .com addresses commanded high prices. It later sold off Network Solutions in a public offering, at a total profit one analyst estimated "in excess of $3 billion."[53] That acquisition allows SAIC to form the bridge between the old Tysons Corner economy of Pentagon-oriented research, and the new Dulles Corridor economy based on telecommunications and the Internet. Recent news about the company has it planning to sell shares to the public, which will make many of its upper- and mid-level employees very wealthy.

DYNCORP

The last of this group of contractors is, like SAIC, also unique. Among the firms we have encountered, it is probably the oldest. Today it performs technical and computer services just like its brethren, but it was founded as a low-technology service firm and moved into the systems arena later. It has also been the subject of more media scrutiny than the others.

In 1946 a group of World War II veterans founded an air cargo company called California Eastern Airways, Inc. From the start the military was one of its biggest customers, especially during the Korean War, when it received large contracts to supply U.S. forces across the Pacific. Through the 1950s it acquired smaller air cargo and air services companies, and began to develop a significant amount of overseas business. In 1957, at the time of Sputnik, the company began to move away from air cargo business and toward the more basic science-oriented work that characterized RAND and the operations research centers. It accomplished that shift in part by aggressively buying smaller companies that specialized in IT. To reflect this shift, in 1961 it changed its name to Dynelectron, and later to DynCorp.

The company never abandoned its skills in aviation, however—something that further sets DynCorp apart from other systems firms in the region, and a skill that would periodically bring the company unfavorable publicity. Throughout its history, the federal government continued to be its main, almost sole, customer. With contracts from the State Department, DynCorp provided aircraft support for the so-called war on drugs in Central and South America. The press called the drive to eradicate the drug trade a war, and it certainly had many of the characteristics of a war. But the Defense Department, concentrating first on the Soviet empire and later on the Middle East, did not wish, or was unable, to wage it. Critics charge that the United States should not contract out the waging of a war, but that decision was made by the Congress, not by DynCorp.

Around the time of the name change, DynCorp moved its base of operations from southern California to Virginia, with its headquarters in Reston.[54]

Like SAIC, DynCorp also allowed its employees to accumulate shares, although outsiders could buy shares as well. It never sold shares on the open stock exchanges.

For much of this time, until 1997, the company was led by Dan R. Bannister, who joined the company in 1953. Bannister was succeeded by former PRC executive Paul V. Lombardi, who sought to move the company further into computer-related and systems work.[55] At the end of the Cold War the company began to diversify into civilian and commercial activities. These efforts met with limited success. Before they had a chance to develop, however, the events of September 11, 2001 created a renewed demand for the military-support skills that DynCorp was known for. It prospered in this environment, but also attracted the attention of other, larger firms seeking an entrée into the lucrative homeland security business growing along the Dulles Corridor. DynCorp could have stayed independent and grown into one of the major players, but in December 2002 its directors agreed to be acquired by the Computer Sciences Corporation, whose California origins were previously discussed.[56]

CONCLUSION

From its humble beginnings after Sputnik in 1957 to the post-9/11 environment, federal contracting in northern Virginia had its share of boom and bust cycles. The overall trend, however, has been positive—dramatically so.

Looking back, one can identify several down cycles. The first followed the end of the Apollo program in the early 1970s, a time of a general recession in technology across the country. While that was happening, high interest rates and the beginnings of an antisprawl movement combined to depress the real-estate market in northern Virginia. Some companies sought to apply their systems engineering skills to non-military or non-aerospace fields, especially to solve urban ills. Boeing, Lockheed, and McDonnell-Douglass each branched into the computer field, with Boeing Computer Services opening a suite of offices in Tysons Corner. Some of those efforts were successful, as long as they concentrated on the technical aspects of implementing a system (such as PRC's contract to develop a nationwide 911 dispatching system). But otherwise they were less successful.[57] Boeing's diversification into light rail cars for Boston's subway, for example, was not well liked by

Boston commuters. The Aerospace Corporation, Boeing, and the Jet Propulsion Laboratory together applied systems engineering concepts to develop a revolutionary Personal Rapid Transit system, a demonstration of which the federal government funded in Morgantown, West Virginia. That proved embarrassing when a malfunction during its 1972 public unveiling trapped President Nixon's daughter Tricia in a rail car. (The Morgantown system has proved to be very reliable and safe ever since. A modern variant of it would be perfect to alleviate the traffic congestion in Tysons Corner today.[58]) BDM was one of the companies that developed controls for the Washington Metro, which opened in 1976 (and which has also worked well).

The second downturn occurred during the administration of President Jimmy Carter (1977–1981), who was perceived as being less than generous to Pentagon requests for military hardware. This was short-lived and the attempt by Carter—the outsider from Plains, Georgia—to put a damper on federal contracting was met with the kind of resistance Washington bureaucrats are famous for.[59] An attempt by a Senate committee to determine the number of consultants hired by federal agencies was met with conflicting information, stonewalling, and obfuscation. A major reason no answer was forthcoming was one mentioned by Earle Williams—they did not consider themselves consultants at all, but rather partners with the agencies who let the contracts in mutually solving a problem. It was too late: by 1977 no one could possibly have answered that question, so intertwined was contracting with the ordinary day-to-day business of the government. Carter's successor, Ronald Reagan, effectively ended the internal debate. His increased defense budgets accelerated the growth of Tysons Corner. His 1983 speech outlining a proposal for a missile defense system, dubbed "Star Wars" by the press, led to large increases in contracts with local companies like BDM, who had been looking at missile defense since its days in El Paso in the late 1950s.

A third downturn, also short-lived, began with the end of the Cold War around 1991. As with the previous cycles, local companies sought to diversify into civilian applications, again with limited success. Those events dovetailed with a crisis in the real estate business caused by the collapse of the Savings and Loan industry, which led to high vacancies in the office buildings along the Dulles Toll Road. But this time there was a countering force mitigating the effects of the Cold War's end. That was the growth of the Internet. Companies like AOL and MCI, based along the Toll Road, absorbed whatever slack there was in the local economy, although not always smoothly. That ended with the collapse of the Internet bubble in 2000 and the 2001 terrorist attacks. The next two chapters will discuss these developments in greater detail, but in every case each time the cycle turns downward, federal contracting has come back at a higher level than it was in the previous peak. Regardless of who occupies the White House, federal budgets trend upward, and the desire to contract for services increases.

When BDM moved, around 1970, to the modest office tower on Leesburg Pike, most of Tysons Corner was undeveloped. Maplewood, the Ulfelders' family farmhouse, was still standing off Route 123, though the family had moved out. The land north of Maplewood was still open fields. A son-in-law, Rudolph Seely, took control of the family's holdings after Mr. Ulfelder's death in 1959 and the granting of power-of-attorney by his widow in 1961. Seely partnered with Gerald Halpin and the Westgate Corporation to develop the land. For a while the Westgate Corporation had its headquarters in the old farmhouse, but that proved impractical, and in 1970 it was torn down.

North of the intersection of Routes 7 and 123 was Gantt Hill, the site of the Rotonda apartments. There the National Automobile Dealers Association, and later SAIC, built their headquarters. In the mid-1960s Gantt Hill was mostly open fields bordering a dense woodlot. To its east was a gravel pit where the Tysons II Galleria mall would be built. The gravel pits were owned and operated by the Bles family, the other major landowner in Tysons Corner. Marcus Bles and his wife Alba built a house on Gantt Hill in 1950 and spent much of that decade buying up smaller parcels of land around them. Beginning in the mid-1960s, after the Beltway and Dulles Access Roads were built, they reversed course and began selling off the land in larger parcels to developers. They moved to Loudoun County in 1969, and their house was demolished to make way for Greensboro Drive. Thus both the Ulfelders' and Marcus Bles's homes were torn down within a year of each other. Unlike the Ulfelders' descendants, Bles sold his holdings, using the proceeds to buy land farther west near Dulles Airport and to repeat the process. In 1980 a reporter asked him if had regrets about leaving Tysons Corner. Bles said that he was glad to be away from all the congestion. The

FIGURE 6. I

Rudolph G. Seely, manager of the Ulfelder dairy farm in Tysons Corner, poses with
West★Group developers (and former Atlantic Research Corporation executives)
Gerald Halpin, Charles B. Ewing, Jr., and Thomas F. Nicholson, circa mid-1970s. The
TRW corporate logo is on the building behind Ewing. © 1980, *The Washington Post.*
Photo by Craig Hendon. Reprinted with permission.

reporter then asked if he ever went back to visit. "Only to the banks," he
replied.[1] Bles did make one exception to his policy of selling property: he
kept a small piece in the shadow of the radio tower, now the site of Clyde's,
the Tysons branch of the famous Georgetown restaurant. Clyde's is a well-
known watering hole, although not on the level of the Wagon Wheel or
other Silicon Valley counterparts, where much wheeling and dealing is said
to have taken place. In Tysons Corner that happens more at the Tower Club.
A plaque at the entrance to Clyde's claims that it rests on the sole remaining
parcel of land still owned by the Bles family.[2]

TYSONS CORNER CENTER, 1968

The first operations research groups to settle in Tysons Corner came from
Maryland or from the western states. Tysons Corner could attract executives
who preferred to live in the Maryland suburbs, but as Fairfax County grew it

acquired schools, parks, and other amenities, making such a commute unnec-
essary. The shopping mall at Tysons Corner did draw from Maryland, however, and it continues to do so. Chapter 4 described the role of Frank C.
Kimball, in partnership with H. Max Ammerman, Isadore M. Gudelsky, and
Theodore N. Lerner, in developing that mall.[3] The Fairfax County Board of
Supervisors approved Kimball's request to rezone land for the mall in July
1962; that is, two years before the completion of the Beltway. When it
opened in 1968, Tysons Corner Center was not northern Virginia's largest
mall, but it generated the most profit.[4] That was partly due to its proximity to
the more affluent parts of Montgomery County, Maryland, but it attracted
the wealthier customers from McLean and Great Falls, Virginia as well.
The developers made a special effort to attract upscale merchants and give it a
refined, expensive look. The shopping mall's cultural impact was enormous—recall the rural character of the county that surveyors for the Beltway
encountered. In 1976, the mall secured the New York retailer Bloomingdale's
as a tenant. This, too, had a cultural effect, symbolizing Virginia's arrival into a
world of sophistication that Northerners took for granted.[5]

So as the 1960s came to a close, Tysons Corner was known to Washington
residents as the location of an interesting suburban shopping mall. The rest of
land was little developed, except for the office building at 8027 Leesburg
Pike, a few small-scale commercial buildings, auto dealerships, and the buildings going up in Westgate, beginning with the one housing the Research
Analysis Corporation. Twenty years later, those four square miles were filled
with development, mainly with office towers housing defense-oriented systems firms. By 1988 Tysons Corner Center had expanded from 1.2 million to
2.1 million square feet, adding a second level, multilevel parking, and more
stores and restaurants.[6] Across Route 123 another mall, the Tysons Galleria,
had just opened. Its stores served a more upscale clientele. The Galleria's
developer, Gerald Hines, represented a new generation of mall developers
who extended the successful model of the 1960s by concentrating on the
most affluent customers. Hines had built similar malls, all called Gallerias, in
the suburbs of Atlanta, Houston, and Dallas, and he designed each to prevent
discount retailers or downscale development from locating in or nearby his
mall.[7] (Hines was unable to do anything about one incongruity: the 1950s-
era communications tower, which visitors to the Galleria pass by.)[8]

On the other side of Tysons Corner Center and next to 8027 Leesburg
Pike, a suite of buildings called Fairfax Square opened at about the same time

as the Galleria. As described earlier, it counted Hermes, Gucci, and Tiffany among its tenants. For a time, above those shops were the offices of Informix, a database company, and Teligent, a wireless networking company—though neither survived the collapse of the Internet bubble after 2000. Just as Tysons II was located near the 1950s-era radio tower, these shops contended with gas stations, discount carpet outlets, auto dealers, adult video shops, and down-scale businesses along Leesburg Pike. The Pike certainly does not resemble Fifth Avenue in New York or Rodeo Drive in Beverly Hills. Slowly these tenants are being driven out by escalating property values. The building at 8027, the oldest high-rise in the area, was demolished in 2004 to make way for a new development to house both offices and more retail space.[9]

In the twenty-year period from 1968 to 1988 the area went through several transformations. The first, well under way by 1968, was the conversion of dairy farms and gravel pits to suburban housing, accompanied by retail shopping malls. That was followed by the replacement of retail by ever more upscale and sophisticated shopping. A third transformation was the influx of office buildings housing military contractors and other high-technology firms. Like the shopping malls, these too followed a progression of occupying ever more expensive and upscale buildings, leading to the demolishing of their first offices. Their expansion also replaced the early retail and service businesses. And in at least one case, this development replaced every house in the Hollinswood suburban housing tract.[10] Some of that transformation was inevitable, a reflection of the move to suburbia common to all American cities in those years. But the introduction of offices for high-technology contractors was not a chance happening. We have seen the role that the West*Group played in luring some of the first systems firms to Tysons Corner. We shall also see that Earle Williams, the president of BDM during this time, played an active role in concert with real estate developers and with county government officials to bring commercial activity to the area.

That transformation did not happen unopposed. By 1970 a reaction to suburban sprawl was taking root. Some of the issues were specific to Fairfax County's rapid suburbanization; others were part of a nationwide environmental movement, which held its first Earth Day that spring. The slow-growth advocates proposed a "pause for planning" to try to control the county's growth. Fairfax County had some degree of planning, but far less than its Maryland counterparts. A regional planning group had developed the concept of "wedges and corridors": a pattern of heavy development

along radial highways coming out of the District, separated by green corridors of parks and environmental conservation areas.[11] That worked well for
Maryland, where streams like Rock Creek and Northwest Branch flowed
toward the District, but in Virginia the streams flowed perpendicular to the
radial highways. Related to the failure of that concept was the hodgepodge
construction of water and sewage facilities in Virginia, with at least one local
stream tasked with the twin jobs of carrying away treated sewage and supplying drinking water from intakes located below the sewage pipes.[12]

That latter situation led to an attempt to slow new construction by impos-
ing a sewer moratorium—no construction was allowed to proceed until it
was demonstrated that the county's sewer capacities were adequate. In 1971, a
county election led to led to the installation of several county supervisors
who could be characterized as slow growth advocates. Chief among them
was Audrey Moore, who represented the Annandale district, in the southern
part of the county. These supervisors did not see themselves as being enemies
of developers, however. They hoped to implement a plan in cooperation with
developers, which they called "Planning and Land Use System," or PLUS.[13]
Developers did not respond favorably to these actions and fought them in the
courts. Almost without exception the courts agreed that the Board's actions
were unlawfully depriving landowners of their constitutional property ownership rights. The newspapers played up the rivalry between Audrey Moore
and Til Hazel, the Harvard-trained lawyer who in those years never lost a
court challenge over zoning. Hazel was known for his good manners, but he
fought for developers' rights tenaciously. PLUS was adopted in 1975, but its
long-term impact on growth in the county is questionable.[14] Although the
consensus in the region is that Audrey Moore's efforts were misguided, she
did enjoy a twenty-year tenure on the Board of Supervisors, serving the last
four as Chairman. Many voters agreed with her view.

THE NOMAN COLE REPORT, 1976

In 1976, after four years of struggles between the Board and developers, the
voters reversed course. That year a Republican member of the Board of
Supervisors, John F. "Jack" Herrity, was elected chairman. That may not seem
like much today, but in the early 1970s Virginia was still part of the Solid
South, loyal to the Democratic Party, although that allegiance was fading.
Recent years have seen a reversal, as Virginia consistently votes Republican

in national elections, although in politics nothing lasts forever. In any event Herrity was not an advocate of slow growth. He fully supported the interests of developers and immediately set out to release the brakes set by the Board.[15] He commissioned a so-called Blue Ribbon Panel, chaired by Noman Cole, and charged it with examining the County's current growing pains and creating a set of practical recommendations for addressing them.

Sometimes called the "Noman Cole Report" after its chair, the committee's product was officially a report of the "Committee to Study the Means of Encouraging Industrial Development in Fairfax County," and was dated June 1976. Among the sixteen members were Earle Williams, John T. Hazel, and Gerald Halpin. The report pulled no punches: it said at the outset that there was a perception that the county was hostile to business, and among the reasons were the sewer moratorium and the pause for planning that the previous Board of Supervisors had instituted.[16] The report listed a number of problems along with recommendations for their solution, but chief among them was the county's increasing reliance on residential property taxes as a source of revenue, rather than what the committee considered a healthy balance of residential, commercial, and industrial taxation. As Earle Williams liked to point out, an office building full of professional workers generates more tax revenue and puts fewer demands on the county than single-family housing built on that same parcel of land. Above all, building an office park does not trigger a need for schools, for whose construction and staffing the county must pay. The report also noted that support for business had to extend beyond helping those business already in the county; its leadership had also to go out and solicit business from other locations around the country.[17]

These and other similar recommendations, especially the last one, were considered radical at the time. But they were implemented. Williams and Hazel both believe that the Noman Cole Report set the pattern for the county's development and prosperity to the end of the millennium. Williams in particular, given his position as president of BDM, was the most active in soliciting business to relocate from outside the region. He joined the County's Economic Development Authority around that time, and he argued for a large increase in its marketing budget. He recalls with a chuckle how they "went ballistic" when he bought a large ad in the *Wall Street Journal*, the cost of which was an order of magnitude larger than anything that group had ever done. One critic said it was a waste of money because "no one reads the

Wall Street Journal."[18] Like the outdoor plumbing Beltway engineers found when surveying, that quote indicates what life was like in Fairfax County at the time. Williams stated that if given the opportunity, he would increase the percentage on tax revenues paid by businesses from the then-current 14 percent to 25 percent. He said that Audrey Moore replied that was a "mathematical impossibility," but by 1986 the county exceeded that goal.[19] Williams played such an active role in these efforts that many attendees at local meetings assumed he was a developer, and did not even realize at first that he was the head of a major local technology-oriented business. His efforts are therefore crucial in addressing the main question of this study: whether the increase in the size and scope of the federal government would have transformed northern Virginia anyway, regardless of what the locals wanted or did not want.

As described in previous chapters, the Tysons Corner area began attracting operations research firms and the branch offices of major defense organizations from all over the country, reaching a critical mass by the mid-1980s. By their nature, they did not manufacture hardware; thus their buildings had no smokestacks, did not pollute the groundwater, and did not they require trainloads of coal or chemicals to operate. As president of BDM, Earle Williams might have not wanted to attract potential competitors to his back yard. But he felt that the Pentagon was in such need of technical expertise that these companies were going to relocate to the Washington area anyway, and he would rather have them as neighbors. By creating a critical mass of such companies, well-educated employees and their spouses from other parts of the country would be more willing to move to Virginia than if BDM were there alone.[20] This perception is thus in alignment with the numerous studies of Silicon Valley, which stress the positive feedback cycle when like-minded firms congregate near one another.

The taxes these companies paid provided a steady flow of revenue to the rest of Virginia, which did not go unnoticed in Richmond.[21] To those who remembered Lee's surrender at Appomattox, that part of the Commonwealth north of the Rappahannock was still "Occupied Virginia." But the state government in Richmond gladly collected the taxes; and not all of those tax revenues were sent back. By 1999, the Fairfax County Chamber of Commerce noted that the county had 14 percent of Virginia's population, but provided 24 percent of the Commonwealth's income tax revenue.[22] If one added the revenues of Arlington County and the city of Alexandria, also areas occupied by the Union army, that percentage of revenue sent to Richmond would be as much as 50 percent, according to some estimates.[23]

Looking back on those years between 1975 and 1990, one can see that developers like Hazel provided the quality housing, schools, and other amenities for skilled workers, the West*Group provided the appropriate office space, and Williams brought about an intellectual and economic environment that attracted these people and their families. Later on BDM saw the downside of that policy, as a severe labor shortage developed, especially for workers with security clearances. At that point the companies could hire only by raiding one another, or by the bigger companies buying out the smaller ones, which amounted to the same thing.

Jack Herrity also played a role. His support brought Fairfax County into the same league as its counterparts in Maryland and the inner suburbs of New York, Chicago, and Philadelphia. This was a time when New York-based corporations were moving their headquarters from Manhattan to suburban Westchester County, New York, and Fairfield County, Connecticut. An office park in Bethesda, Maryland, also near the Beltway, was already attracting such tenants. Some movement of private corporations to the Washington suburbs was inevitable, but not necessarily to Tysons Corner. For all its prescience, the Noman Cole Report made only a passing mention of Tysons Corner, which would become the true commercial center of the county, and indeed the whole region outside of the federal core. Williams was not a developer, although he recalled with amusement how people labeled him as such. Halpin, who was a developer, and whose company concentrated on Tysons Corner, saw Jack Herrity as less influential. A 2003 newspaper story about Herrity echoed that sentiment. It quoted another developer saying that presidents Jimmy Carter and Ronald Reagan, whose defense policies channeled large contracts to private contractors in Tysons Corner, were the real engine driving Fairfax County's tranformation.[24] As previously mentioned, Carter and his staff made an attempt to rein in the growth of private contractors and consultants, but they failed. A quarter-century later, President Carter's support seems parsimonious only in contrast to the gushing money flowing from his successors.

HIGHWAYS

Another issue that the Noman Cole Report did mention, and that Herrity took a personal role in addressing, was the need for better transportation.

The aerial photograph of Tysons Corner in the mid-1960s (figure 4.1) shows an excellent network of highways in anticipation of development, but by the mid-1970s this situation had been reversed. The report noted three "bottle-necks at key urban center locations (Tysons Corner, etc.), (2) Lack of new cross-county arteries, [and] (3) Lack of adequate access to the Dulles Airport business area."[25] The Dulles Access Road was intended to provide access from the District, but between Tysons Corner and the Potomac River the route was contested. A Potomac River bridge at Three Sisters Islands was to con-nect to an inner Beltway in D.C., but the residents of Georgetown and Glover Park stopped the planned I-266. The Three Sisters Bridge was can-celled in 1972. Likewise, Interstate 66 was built westward to the Shenandoah Valley, but it, too was stalled inside the Beltway. Herrity became chairman of the Board of Supervisors in 1976 and immediately made the completion of this latter road a high priority. He faced strong opposition from the residents of Arlington, who were well organized like their counterparts in the District and Montgomery County, Maryland. The road's advocates succeeded in bringing the interstate system into the District by tying support for I-66 to support for an extension of the planned Metro rail system, eventually opened to Vienna in 1986.

After a long struggle, President Gerald Ford's Secretary of Transportation, William Coleman, approved the road in 1977 after Herrity reversed the Fairfax Board of Supervisors' earlier opposition. The ten-mile extension of I-66 opened in December 1982, to a Potomac crossing a mile downstream of the Three Sisters Islands. Part of the right-of-way came from an abandoned branch of the Washington and Old Dominion Railroad. To get the road built, Coleman, Herrity, Virginia Governor Mills E. Godwin, Jr., and other advo-cates had to make compromises. I-66 thus acquired properties that make it unique in the nation's interstate highway system. Other interstates may have some of these features, but none has them all.[26] These are

· No large trucks at any time.
· Only two travel lanes in each direction, with an informal agreement that it will not be widened (just who made this agreement, and with whom, is contested, and it is a surprise that it has lasted as long as it has).
· A bicycle path (the Martha Custis Trail) parallel to I-66 from the Potomac to Falls Church.
· A provision for the Metro rail line in the median west of Ballston.

• A High-Occupancy Vehicle (HOV) restriction—in other words, prohibiting a sole occupant—for large portions of the day. Initially it was HOV-4, and was later reduced to HOV-2. This number, and whether an infant in a child seat counts as a person, is likewise debated from time to time.

• Noise barriers. These are now common throughout the region, but this stretch of I-66 pioneered the concept. Through Arlington, much of the highway is depressed to further minimize its visual and acoustic impact.

Interstate 66 came near but did not pass through Tysons Corner, which by 1982 had become the commercial center of the region. Two other roads completed in the 1980s had a more direct impact on Tysons Corner. In reverse order, they were the connection between the I-66 and the Dulles Access Road at Tysons Corner, opened in 1985, and a toll road parallel to the Dulles Access road from Tysons Corner to Dulles, opened in 1984.

The connection from the Dulles Access Road to I-66 was only 2.5 miles long but was the missing link in the whole plan to make Dulles the region's premier airport. The interim solution, funneling traffic onto the Beltway or Chain Bridge Road, was awkward. The gap was one reason Dulles traffic was so sparse for many years, and it accounted in part for the ascendancy of Tysons Corner over the other Beltway interchanges in Virginia. By the time the connection to Washington was completed, Tysons was self-sustaining and needed no artificial stimulant. Thus Tysons Corner benefited from the opening of the connection, just as in an earlier day it had benefited from its absence.

General Quesada, who brought Dulles Airport into existence, had from the start reserved land for a commercial highway parallel to the Dulles Access Road. But its completion was delayed for over twenty years, until a method of financing it through tolls was adopted. When it opened it was an immediate success. Its biggest impact was on Reston, which previously had no access to or from the Dulles Access Road, despite straddling it. Within a few years, Reston's skyline took on the characteristics of a bustling city. Robert E. Simon's initial vision for the town finally became a reality. The Dulles Toll Road, later named the Hirst-Brault Expressway after two local politicians, also drove development farther west, through Herndon, Sterling, and beyond. This sprawl stands in stark and unfortunate contrast to the planned city of Reston.

One more road needs to be mentioned. It was not built until the turn of the millennium, after Herrity was no longer on the Board of Supervisors, but it is named after him and reveals another facet of the region's growth. Recall that the initial plans for Washington included not just "the" Beltway but a series of circumferential freeways around the city. Only one was built. The Inner Loop was stopped by local opposition, and beltways further out were stalled in the early planning stages. The Virginia portion of an outer beltway was, however, partially built and has been opening in pieces as segments are finished. It is not built to Interstate standards and has a few traffic signals. It was called the Fairfax County Parkway; it was recently renamed the 'John (Jack) Herrity Parkway'. It has been especially beneficial to George Mason University, located at the south side of the city of Fairfax, which now has direct highway access to the booming Dulles corridor.

Developers want to extend this road to the north, crossing the Potomac near Blockhouse Point and connecting with I-370 in Rockville, Maryland.[27] They call the connection the Techway: connecting the Internet and telecommunications centers of Reston and Herndon with the biotechnology centers in Rockville. If combined with a planned extension of I-370 in Maryland to I-95, this extension would form two-thirds of a circle of the original Outer Beltway. There is strenuous opposition at nearly every segment of this planned road, especially from residents who live near Blockhouse Point, one of the most scenic places on the Potomac. It is fitting that this road be named after Herrity, but the road is also a reminder that the initial highway plans of the 1960s were only partially fulfilled. Future generations will pass judgment on this turn of events.

GEORGE MASON UNIVERSITY

The shelves of books about the growth of Silicon Valley and Route 128 all cite the need for a research university allowing a constant flow of professors, students, and ideas from the academy to industries. MIT's role in the development of Route 128 is without question. In Silicon Valley the story is more complex, as the founders of the pioneering companies like Shockley Semiconductor, Fairchild, and Intel did not come from Stanford, though Stanford's role was significant.[28] Northern Virginia's story is different than both of these. The major university associated with this region, George Mason University in Fairfax, grew up with the operations research and systems companies, in a symbiotic relationship. The university was named after George Mason (1725–1792), a Fairfax County resident, Virginia delegate to the Constitutional Convention, and author of the Virginia Declaration of Rights, the predecessor to the U.S. Constitution's Bill of Rights. The school was founded in 1957 as a two-year extension of the University of Virginia and was located in a former elementary school at Bailey's Crossroads. It became a four-year school and moved to a campus near the city of Fairfax in the mid-1960s.[29] It was granted separate university status in 1972; by then it had an enrollment of about 4,000 students.

In 1978 George Johnson, an English professor and dean at Temple University, was hired as president, remaining at that post until 1996. He thus arrived about two years after Jack Herrity became chairman of the Board of Supervisors, and after the Noman Cole Report was issued to outline the region's future. Johnson sensed that the region was poised for dramatic growth, based on the defense contractors and systems companies then moving into Tysons Corner and the surrounding areas. He knew also that the university had little clout with Virginia's politicians or the state's financial institutions—the latter called "Main Street" after their location in Richmond. The people with influence in Virginia were typically alumni of the University of Virginia, founded by Thomas Jefferson, or of William and Mary, the second-oldest school in the country and Jefferson's own alma mater. To them, George Mason would never be considered on an equal footing. And because it was so new there was little opportunity for its graduates to find their ways into the Richmond establishments of power that allocated resources to universities.

At the same time, Johnson saw that the University of Virginia, as good as it was, lacked the size and diversity that a major university serving Virginia ought to have. The University of Virginia had about 12,500 students; by Johnson's reckoning it should have had about twice that.[30] Virginia's other large state university, Virginia Tech, was large and diverse but located in Blacksburg, 250 miles to the west in an isolated mountain region. Across the Potomac was the University of Maryland. This was a large school, with a full range of departments and a diverse student body. It was strong in computer science, physics, electrical and aeronautical engineering—all necessary to support high-technology industries. But Maryland failed to develop the kind of synergy with entrepreneurial companies that Stanford or MIT did—neither in College Park, where it was located, nor in the surrounding region. Maryland's failure was in part due to the simple advantage of the Pentagon's location in Arlington and the small number of high-speed highway Potomac crossings. Johnson also felt that Maryland failed to seize an opportunity to reach outside the campus when this phenomenon was beginning. Johnson moved into the gap and rapidly expanded the university to an undergraduate enrollment of approximately 16,000 by 2002, below Virginia Tech's 22,000 but above the University of Virginia's.[31] The studies of Silicon Valley show that the Valley was helped by the presence of Stanford University, but the example of the University of Maryland suggests that a strong research university alone is not sufficient to trigger such growth.

The new president of George Mason University accomplished this transformation by reaching out to the systems engineering companies he saw growing rapidly in the county. Johnson remarked that he wanted to make the university a place with Harvard aspirations, but connected to the region like a community college.[32] He not only visited the local Chamber of Commerce; he joined it, becoming its vice-president. "My friends at Temple never would have believed it," he said, but the local business community welcomed him. He spoke with Til Hazel and Earle Williams almost daily; he also worked with John Toups of Planning Research Corporation and the CEOs of most of the other Tysons Corner systems firms. In their day-to-day work those people competed with each other, but at meetings on the George Mason campus they found common ground and threw their support behind Johnson's vision. Their support was twofold: gaining the financial support of Richmond's Main Street, and recruiting high-quality engineering faculty to come to what was then a little-known school.

Johnson, in return, focused the growth of the school on three areas, one of which we now call IT. Perhaps because of his liberal arts background, or perhaps because of his vision of an upscale community college, IT evolved as a more practically focused field than computer science, from which it descended. That dovetailed nicely with the needs of the Tysons Corner companies while further setting George Mason apart from its university counterparts elsewhere. Johnson described a visit with James Fletcher, the administrator of NASA at the time and a respected engineer, whom he asked for advice on how to build up the university's engineering reputation. Fletcher responded that George Mason needed to establish basic competency in mechanical, civil, electrical, and chemical engineering before proceeding. Johnson rejected that advice, and chose to focus on the information sciences and related engineering.[33] Universities are by nature conservative institutions, and many to this day resent the way George Mason has grown. It still lacks the political and financial clout that the University of

Virginia enjoys, mainly because it is not old enough to build up a cohort of
wealthy and influential alumni. It has no big-time football team, which is
both an asset and a liability.[34] Football games attract wealthy alumni back to
the campus, but the correlation between a school's standing in football and
its academic reputation is a matter of dispute. Some Marylanders, for example, feel that the school's academic reputation suffers as a result of its football
program, especially on nights when the students get rowdy after a victory.
GMU is still operating like a community college, albeit one with two Nobel
laureates on its faculty—the first and only Virginia university to have two
Nobel laureates. At the turn of the century, other universities have begun to
adopt at least the rhetoric of GMU's philosophy, if they do not always implement it fully. Satellite "campuses" (often no more than leased office space) of
Virginia Tech, Johns Hopkins, George Washington University, and other
regional universities are sprouting up in office parks all along the Dulles Toll
Road. Each hopes to capture the synergy that GMU developed with the
local technology community. They are a threat to GMU's stature, but for
now only a minor one.[35] For the 1980s and 1990s, at least, the strategy
worked for the northern Virginia economy.

STAR WARS

BDM's first contracts for the Army at White Sands, New Mexico included
studies of the NIKE-X antiballistic missile system. Throughout the history of
that company, from its founding in 1959 to its merger with TRW in 1997,
missile defense remained one of BDM's core skills and a major revenue
source. The complexities of detecting, intercepting, and destroying an incoming ballistic missile, which may be carrying a nuclear warhead, are immense. It
is precisely the kind of problem that the mathematical techniques of operations research and systems engineering were developed to solve. In fact, ballistic missile defense is a natural descendant of the defense of the British Isles
from German air attack during World War II, which led to the development
of OR in the first place.

The NIKE-X antimissile system, tested in the New Mexico desert, was
one of many the United States developed since the onset of the Cold War.
But none proceeded to a point where they were considered effective. In the
1970s the Soviet Union accelerated its production of nuclear warheads to a
point where in sheer numbers they overwhelmed the United States' ICBM

arsenal, leading some American analysts to fear that the Soviets would be tempted to mount a first strike. The American response was to develop a new ICBM, the MX or "Peacekeeper," using a silo configuration that would allow enough missiles to survive a first strike. Although the MX might give the Soviets pause, it did not address the root of the problem—namely, the threat to the United States from attack by the ballistic missiles that could not be stopped.[36] The Soviet buildup of warheads was not as dramatic a threat as Sputnik was in 1957, but it was just as real, and even more disturbing to U.S. military planners who knew what it implied.

Although the Carter administration worked diligently on this problem, the perception that Carter was not doing enough was at least one factor leading to Ronald Reagan's November 1980 election. But Reagan's options were limited. He was reluctant to construct large ICBM bases in Nevada and Utah—the residents of these states were among his strongest supporters, and they opposed such bases.[37] Before his election he visited U.S. Air Force facilities at Cheyenne Mountain, Colorado and was briefed on the details of the U.S.-Soviet arms race. The Air Force officers leading the tour told him, among other things, that the Cheyenne Mountain facilities did little to reduce the ability of the Soviet Union to attack and destroy American cities with nuclear weapons. The visit, plus the frustrating debates over the MX, led Reagan to question the policy of mutual assured destruction that defined U.S.-Soviet relations for decades. Early in his term as President, proponents of ballistic missile defense saw an opportunity to press their case. The technical problems of intercepting and destroying an incoming warhead had not been solved, but there had been advances that showed promise. After some initial setbacks, the efforts to reach the President succeeded and were dramatically revealed to the public in a speech by Reagan in 1983. On March 23, 1983, he said, "I call upon the scientific community who gave us nuclear weapons to turn their great talents to the cause of mankind and world peace; to give us the means of making those weapons obsolete."[38] The press quickly dubbed his speech the "Star Wars" speech, after the blockbuster movie (no pun intended) of that name.[39] The official name of the system Reagan was proposing was the Strategic Defense Initiative (SDI). Its proponents never liked the other term, and they still do not.

The immediate impact of this proposal was a dramatic increase in research and development funds for this system. The impetus for the speech came from ideas developed by Edward Teller at the Lawrence Livermore Laboratory, but in 1983 the concept was very much in the preliminary stages. A provision of a 1972 treaty with the Soviet Union had restricted the actual deployment of a missile defense system; therefore whatever monies were allocated for SDI had to go for research, unless that treaty were amended.[40] Between its formal inception in 1984 and 1987 the program spent over $7 billion, with projected costs of $100 billion to reach the deployment stage. And it was not just the aerospace giants like Lockheed or Rockwell that got these contracts. According the Federation of American Scientists, SDI produced a "gusher" of money that was distributed to over 500 for-profit companies.[41] For the first two years of the Reagan presidency the press wrote endlessly about his desire to shrink the federal budget, using his point man, Budget Director David Stockman. For the Washington, D.C. region, those efforts were more than offset by increases in the defense budget, especially after 1983.

None of the top twenty companies receiving SDI contracts in 1987 were located in northern Virginia, according to the Federation of American Scientists. But the Tysons Corner firms did enter into subcontracts with the major aerospace companies. BDM's long history of involvement with this missile defense positioned it to take advantage of the new contracts being offered. Annual reports from 1984 through 1987 show that SDI contracts were a major source of the company's revenue. Its 1984 report listed the army's High Energy Laser System Test Facility, the High-Endoatmospheric Defense Interceptor (within the Earth's atmosphere), and the Exoatmospheric Re-entry Vehicle Interceptor System (in space) as among the SDI projects it was supporting. BDM provided systems analysis, integration, testing, and logistical support.[42] The following years saw a surge in revenue for the company, with more contracts for other aspects of SDI. BDM's 1986 annual report devoted four pages to a description of its work on SDI, as well as the computing spin-off of SDI, the Strategic Computing Initiative, intended in part to develop a computation capability to handle the needs of missile defense. Looking over these reports from that decade, we see that BDM developed a special expertise in command-and-control facilities and the integration of computers and communications devices (C^3I), which enabled it to win large contracts for SDI. Other Tysons Corner firms, especially the MITRE Corporation, also got this work.[43]

The impact of SDI was twofold. The first was local. Tysons Corner, already supercharged with defense contracts and known for its shopping malls, became even more prosperous and congested. The toll road to Dulles and the I-66 connector to the District opened up, but they were clogged with traffic almost from the day they opened. The Beltway was also widened at this time, and new lanes were added to the American Legion Bridge crossing the Potomac from Montgomery County. The failure of these roads to alleviate congestion surprised transportation planners, but was a natural consequence of what was developing at the Pentagon. The opening of the toll road led some of this defense work to shift westward to Reston. Office rents in Tysons Corner increased dramatically, further pushing development along the toll road, but vacancy rates in Tysons Corner remained low. The low rents of a few dollars a square foot attracted tenants to Tysons Corner in the early days. By 1990 the rents were over $10 a square foot, but it did not matter. The citizens of Fairfax County were of two minds about this growth: they welcomed the new jobs but balked at the congestion, and they suffered from sticker shock when faced with the prospect of buying a home in the area. Audrey Moore, the scourge of Til Hazel, George Johnson, and Earle Williams, was not only reelected to the Board of Supervisors, she was elected its chair in 1987. She made a last attempt to control growth, but it was in vain. In 1991 during a brief recession, she was defeated by Tom Davis, who later went on to represent the region in Congress. J. Hamilton Lambert, who was County Executive at the time, remarked that "I'd rather be stuck in traffic on my way to work, than to be driving down an empty highway looking for a job."[44] Slow growth advocates did not agree but were losing the battle. They shifted their efforts to neighboring Loudoun County to the west, which since 1987 has been undergoing a similar transformation from farming to a dense suburban Edgeless City; whether it does so in a controlled fashion remains to be seen.

The other effect of SDI was more profound. Its proponents claim that it led to the collapse of the Soviet Union and the end of the Cold War. This is a controversial point, one that many dispute. SDI was never successfully tested. Its champions reply that it forced the Soviets to counter with a missile defense system of their own, which they were unable to afford. It also made them realize that the United States was not going to accept the numerical advantage they had developed in ICBMs. Whether SDI did in fact cause the Cold War to end is beyond the scope of this study. By 1991 the Berlin Wall had fallen and the Soviet Union was gone. No exchange of nuclear weapons preceded those events; no atomic bombs have been dropped in anger since August, 1945.

THE LAST SUPPER

For the U.S. aerospace and defense industries, the effect of the end of the Cold War was similar to what happened in 1945 after the Japanese surrender. From 1988 to 1995 defense spending dropped about 27 percent in constant dollars, according to one estimate.[45] Some of this reduction had been anticipated, but it did throw the aerospace industry into turmoil. In 1993, President Clinton's Deputy Secretary of Defense, William Perry, held a dinner at his home in McLean, to which he invited a number of industry executives. At the dinner, since known as the "Last Supper," he forecast that future defense spending would only be enough to support about half the number of companies then in existence. His remarks implied two things. First, the companies represented at the dinner had to combine with one another or go out of existence. Second, although Perry had no authority to say so, if these companies chose to combine, they would not run afoul of antitrust laws. It turned out that not only were there no antitrust concerns, the Defense Department even helped pay for the costs of some of the mergers.[46]

From the over fifty companies that supplied satellites, aircraft, missiles, and other defense hardware in the early 1980s only five large defense suppliers remained by the turn of the century: Boeing, Lockheed Martin, Northrop Grumman, General Dynamics, and Raytheon.[47] Gone were many familiar names of aircraft, spacecraft, and computer systems: McDonnell Douglas, North American, Rockwell, Litton, IBM Federal Systems, Fairchild, General Electric Aerospace, Aerojet, TRW, Hughes, and many others.

For the companies in Tysons Corner, this process was already underway by the time of Perry's dinner. It accelerated through the 1990s and into the next century. BDM had been bought and sold twice and was part of TRW by 1997. TRW was bought by Northrop Grumman. PRC merged with Litton in 1995; the combined company was then bought by Northrop Grumman. BTG was bought by Titan in 2002, and DynCorp by Computer Sciences Corporation in 2003. Five years into the new millennium, the biggest employers in northern Virginia are Northrop Grumman, Computer Sciences Corporation, and SAIC. General Dynamics has less of a presence, even with its headquarters in Falls Church. In 2004 it had about 5,000 local employees, compared to Northrop Grumman's 19,000 and SAIC's 15,000.[48]

Northrop Grumman's dominance in northern Virginia stems from its acquisition of the descendants of PRC and BDM, the two companies that helped define Tysons Corner. It does not build airplanes in the region. Indeed, the company builds few piloted aircraft anywhere.[49] That may disappoint those who remember the beautiful airplanes Northrop used to build, or Grumman's Lunar Modules, which took twelve astronauts to the moon between 1969 and 1972. These beloved craft are the centerpieces of exhibits at the National Air and Space Museum and elsewhere.[50] Ronald D. Sugar, Northrop Grumman's CEO in 2005, does not share that nostalgia. To him, "We fundamentally are an information and electronics company that also happens to build aircraft, ships, and spacecraft."[51] Northrop Grumman's employment advertisements in local papers seldom show flight hardware but stress its expertise in systems integration, command-and-control, and that ubiquitous acronym, IT. Boeing and Lockheed Martin build more flight hardware than Northrop Grumman, but they derive a lot of their revenue from IT as well. Lockheed Martin is "the number one provider of information technology to the U.S. government."[52]

This consolidation did not hurt the local economy. The systems analysis work was still being done, and the cuts in defense spending did not last long. Defense cuts were offset locally by increases in the budgets of the intelligence agencies, which are not part of the DoD and whose budgets are not revealed to the public. At the same time Pentagon contracts got a lot bigger, so that instead of letting dozens of contracts in the one-half to one million dollar range, the Pentagon now advertises for one contract in the hundred-million dollar range or higher.[53] The total amount of money grew, but a small company could not bid for these kinds of contracts. In earlier years there was an attempt to separate the analysis and evaluation of a weapons system from the companies who made its hardware, but after the industry's consolidation this is no longer practical. Whether that leads to abuses of contracting is an open question. One fundamental result of this consolidation is that the Tysons firms no longer serve as the buffer, or interface, between the major suppliers of hardware and the Pentagon. They are now all divisions of those hardware suppliers. The Pentagon will have to look elsewhere for such advice, or else do without that advice at its peril.

Another major effect of these events is that, while defense contracting is as vibrant as ever in Tysons Corner and the Dulles Corridor, the companies now have their headquarters in southern California, where the headquarters of Northrop Grumman, SAIC, and CSC are located. Through the mid-1990s, Tysons Corner grew as a result of close relationships among local politicians, developers, and executives, including Johnson, Herrity, Hazel, Williams, and Toups. George Johnson recalled that whenever he spoke to reporters as the president of George Mason University, they could assume that he was speaking for Earle Williams and Til Hazel, so close were their interests.[54] A tenet of business school wisdom is that the CEO should focus on the business and not be distracted by charity balls, local politics, and other externalities. Earle Williams violated this rule with impunity. As he steered his company through double-digit growth rates year after year, he found the time to work with the Chamber of Commerce, the Board of Supervisors (including Audrey Moore), local civic organizations, and local charities.

That kind of alliance will not repeat itself. The head of Northrop Grumman will not sit in on local planning board meetings. It may no longer be necessary. Recall how Tysons Corner initially benefited from being at the temporary end of the Dulles Airport Access Road. When that road was finally extended to Washington, Tysons was large and robust enough that it no longer needed such an artificial stimulus. Perhaps now it will continue to thrive without the alliances that were so crucial to its earlier prosperity.

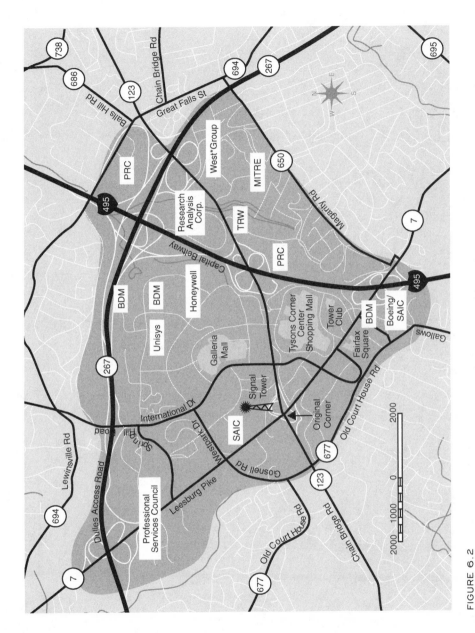

FIGURE 6.2

Map of Tysons Corner, circa. 2000. The approximate locations of BDM and other defense contractors are shown.

Source: ASAP Graphics.

In the nineteenth century, the Washington and Old Dominion Railroad set the pattern of settlement in Fairfax County. The towns of Rosslyn, Vienna, Herndon, Sterling, and Ashburn prospered, while towns off the line remained small or vanished. In the twentieth century, the Beltway, Route 7, and Route 123 set the pattern. Tysons Corner developed rapidly, while other parts of Fairfax County developed more slowly. In the twenty-first century one finds a more complex pattern. Tysons Corner continues to thrive, while the old route of the W&OD, now replicated by a sequence of limited-access highways, is the site of something new. This Dulles Corridor is a center for telecommunications companies, with a concentration on Internet-related activities. A more accurate name for the string of settlements following the old rail line would be "Internet Alley": a string of places where the Internet is managed and governed.

The Internet's rise to its present position in the world economy came suddenly and took many by surprise. Scholars from a variety of disciplines have sought to understand it and chronicle its history. Each approaches the story from a different angle. Some look at technology, focusing on the engineering challenges that the Internet's creators faced. Others look at the mathematical theories that made it possible. Some focus on the Defense Department, for whom the Internet was developed to address communications under wartime conditions. Still others take the opposite tack and see it not as a product of military values, but of 1960s-era countercultural values of collective ownership and participatory democracy. Some see the Internet's creation as less of a technical history than one of the marshalling of political will. In 1999 presidential candidate Al Gore caused a controversy when he discussed his role in the Internet's creation, emphasizing the political dimension over the other dimensions.[1] In chapter 3, we gave credit to President Eisenhower for the creation of the Interstate Highway system. Designing, routing, and building those highways challenged the civil engineering community, but the political challenges are what people remember. For the Internet, the consensus seems to be the opposite: its creation was less a political accomplishment than a matter of computer science and software engineering.

This chapter looks at the Internet in terms of another dimension, geography. It looks at the Internet as it is manifested along the former route of the Washington and Old Dominion Railroad through northern Virginia. The reader may think this point of view is even more absurd than Al Gore's supposed claims. What could a long-abandoned, Civil War–era railroad have to do with the Internet? I do not claim that this is the best way to look at the history, only that it is a valid way. In contrast to some histories, especially the hardware-oriented histories that assume theirs is the only valid approach, this approach does no harm as long as the reader understands why I chose it. Geography always plays a role in technical innovation. It deserves emphasis for no other reason than the frequently heard claim that the Internet has annihilated space. Advances in fiber-optic transmission cables, plus the flatness of the World Wide Web's addressing structure, make any Web site as accessible as any other, regardless of where in the world it is located. And people can shop and even run an online business from home. In this chapter I give a modest rebuttal to these hyperbolic statements in the hopes of revealing that cyberspace, too, has a connection to the earth, and that at least one connection is found along the Dulles Corridor.

How did this Internet Alley emerge? What relationship does it have to the previous commercial development that came from the railroad? And what is its relationship to the military contracting happening in Tysons Corner? By the time Internet Alley emerged, the W&OD railroad had been turned into a popular recreational trail. But the railroad laid the foundation for the rapid growth that was to follow. The decisions locating Dulles Airport at Chantilly and Reston at the Wiehle property were not dependent on the W&OD. But without the W&OD, they would have remained separate islands. The rail line provided a vestigial connection, not only between Dulles and Reston, but also with other towns farther out and with Tysons Corner and Rosslyn closer in.

Among the many responses to Joel Garreau's book on edge cities was a critique by Robert E. Lang of Virginia Tech, who argued that the real story of recent suburban development was not in edge cities but in open-ended

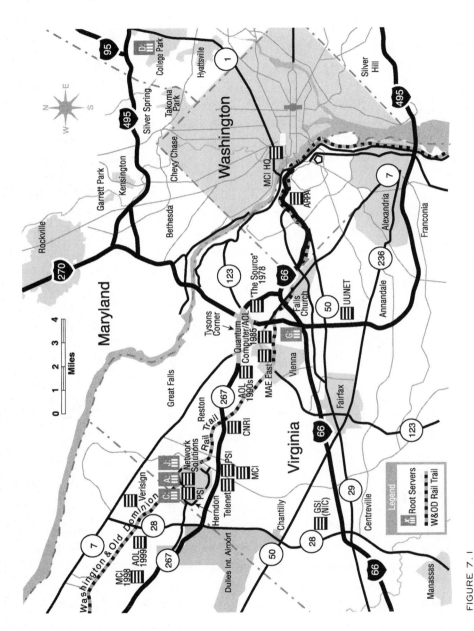

FIGURE 7.1

Internet Alley circa 1990. The most popular term is now "Dulles Corridor," but in fact it now extends well past Dulles Airport to Ashburn to the west, and to Chantilly to the south. *Source:* ASAP Graphics.

sprawl from city centers into the hinterland, with no definable edge at all.[2] Lang titled his study *Edgeless Cities* deliberately to set his conclusions apart from Garreau's. For Lang, Tysons Corner is an anomaly, its sharp boundaries caused by the chance location of greenbelts and residential areas that restricted its spread. He argued that the norm for the twenty-first century is undifferentiated sprawl, with no concentration of commercial, residential, or retail development. The Dulles Corridor fits Lang's model: it stretches over thirty five miles in length and is poised to do what the railroad failed to do— cross the Blue Ridge and spread into West Virginia.[3] By 2005, development was most evident along the W&OD rail-trail between Sterling and Leesburg, where heavy construction trucks lumber down old two-lane farm roads, throwing up huge clouds of dust as they shared the lanes with commuters on their way to new office parks. The new name for this area is "Dulles," which its inhabitants prefer over the older Sterling and Ashburn. A second spur of development is spreading south to Chantilly, where a cluster of Homeland Security agencies and contractors have taken root along with the National Air and Space Museum's spectacular new annex building, the Steven F. Udvar-Hazy Center. Neither spur has a recognizable edge. Within these zones one finds large, gleaming new office parks, pockets of open land, gravel pits, dense clusters of residential townhouses, strip malls, light industrial buildings, and even a small farm or two. There is no pattern or logic to their location.

A corridor this long will obviously contain a wide range of activities. Military and national security contractors and related government agencies are found all through these corridors. Earlier chapters mentioned - several, including the Center for Innovative Technology, the National Reconnaissance Office, DynCorp, and CACI. There are many others as well, which might lead one to conclude that this corridor simply carries the spillover from Tysons Corner. But this corridor grew from different roots. Tysons Corner firms descended from the operations research analysts coming from the Pentagon and RAND. The Internet firms grew in part from military contracting, but also from computer and telecommunications services companies.

Some Dulles Corridor firms began as computer services companies that supplied data processing to a variety of government agencies. One entrepreneur called this "government business" to differentiate it from "government contracting." To him, government business means the government agency is a customer, not a client, and it is buying mainly computer services.[4] Firms in this category served the Defense Department, but did as much business with the Department of Justice, the IRS, the Department of Health and Human Services, and other civilian agencies. Veterans of those companies' pioneering years frequently mention the Department of Health, Education, and Welfare as a major customer, one whose budgets were on par with the Defense Department's. The establishment of Medicare in 1965, part of President Lyndon B. Johnson's "Great Society," was for these computer service companies what Sputnik was for defense contractors. And unlike defense-oriented contractors, these companies saw no great advantage in a Virginia location.

Like their Tysons Corner counterparts, computer services firms required skilled workers with systems engineering skills. Retired military officers were found among their employees, but their knowledge of the Pentagon's Corridors was less critical. Security clearances were not as important. Proximity to the Pentagon did not matter at all for those serving the Department of Health and Human Services, located in Maryland and the District. That explains at least part of the reason for the sprawling nature of Internet Alley. Whereas many Tysons firms moved there from southern California, many of the computer services firms moved to the Dulles Corridor from the District, especially from Wisconsin Avenue near the National Cathedral.[5]

The founders of BDM recalled that they could not afford a mainframe computer in their early years. For most of their work they relied on a desktop electronic calculator. If they needed more computing power they rented mainframe time. The common perception of time is that it flows independently of human intervention or even human awareness, but for computer vendors time was a real quantity that could be bought and sold. BDM and its neighbors were never happy with this arrangement, but they had little choice, at least until low-cost minicomputers became available around 1970.[6] Through the 1960s and into the 1970s, IBM mainframes ruled this arena. Its computers were powerful, large, and expensive. They required a climate-controlled room with a false floor, a dedicated power supply, and full-time technicians to run them. They were programmed by punched cards, in batches: one submitted a program and data on a deck of cards, and retrieved the answers some time later on after that program was run, alongside other users, programs. This arrangement was not the product of some diabolical plot by computer vendors to inconvenience users;

it was a product of the economics of computer processing. Mainframe computers required specialized physical facilities including conditioned power and air. The costs of using them were measured by the time, in fractions of a second, that a computer's central processor worked on a problem.

This economic situation led to the creation of companies that would purchase a large computer and then sell processor cycles to customers who could not afford their own computer. In the simplest cases, these companies would collect decks of punched cards from a client, bring them to their computer, run the job, and deliver the results. A more convenient arrangement would be to have a card reader or other input device connected to the mainframe but located on the customer's premises. That operation was known by jargon like "Remote Job Entry" or "Conversational Remote Batch Entry." It did not overcome the inherent unfriendliness of batch processing, but for those who ran the same program every day with only the data changing, it was effective.

This activity began in the Washington region in 1952, at the dawn of the computer age. That year Herbert W. Robinson, a British scientist who had done operations research for Winston Churchill's scientific adviser during World War II, established a not-for-profit Council for Economic and Industrial Research in Arlington. C-E-I-R, as it was called, initially did OR for the military, like the other centers in Arlington at the time. Robinson later converted C-E-I-R to a for-profit status, and it diverged from the other OR centers by concentrating on computer services. It grew into one of the largest services firms, using a succession of ever more powerful, mostly IBM computers. In 1968 it was bought by Control Data Corporation (CDC) and helped establish CDC's presence in northern Virginia.[7] Another early entry into this field was California-based Computer Sciences Corporation (CSC), which opened a Virginia office in 1962. That was followed by branches elsewhere in the region, setting the stage for its acquisition of DynCorp decades later.

In the mid to late 1960s a second wave of computer services firms followed. Boeing Computer Services was among them, as its parent company was looking to diversify in the wake of a downturn in the commercial aircraft business. Unlike its many other forays into non-aerospace products, such as mass transit cars, Boeing's computer services were successful, with Boeing claiming that by 1975 it was the "nation's fourth-largest computer services company."[8] The Seattle-based company opened an office in Tysons Corner in 1972, along with other offices in the region. In 1977 it consolidated these offices and established a large campus at Boeing Court off Gallows Road, now occupied by SAIC.

Other aerospace companies established Virginia centers for remote computing services. MITRE opened an office in the Westgate office park in 1966; its Massachusetts office continued to concentrate on air defense, while the local offices provided computer services of a more general nature. The California aerospace firm Litton opened a center in Reston at the same time; that set the stage for its later acquisition of PRC. Honeywell, a Twin Cities aerospace and computer company, opened a center in Rosslyn and then in Tysons Corner in 1970.[9] The Tysons Corner building was one of the first high-rise buildings there, and its construction began Tysons' transformation into its present dense, high-rise office pattern. All of these companies had their headquarters elsewhere. One exception was American Management Systems, founded in 1970 by five of Defense Secretary McNamara's "whiz kids" to provide computer services to the government.[10] AMS was founded in Arlington and later moved to Fairfax, near the site of the Civil War Battle of Chantilly.[11] It is no longer independent but retains offices there. In Maryland, General Electric established General Electric Information Services in Bethesda, later moving to Rockville.

One driver for this second wave of computer companies was the evolution of technology. In 1964 IBM introduced a vastly more capable computer, the System/360, which transformed the industry and allowed IBM to dominate the market for decades. The System/360's complexity created a need for systems analysts who could translate a client's problem into programs the computer could solve, and for programmers who could write those programs. In an earlier era, IBM would have supplied most of that software and would have included its cost in the rental of the machine. With System/360 that no longer was the case. The Honeywell, Control Data, and General Electric centers all used computers of their own design, but those machines had comparable complexity, and their customers also had similar needs.

Another technical advance in computer design in the 1960s eventually had a greater impact. The Remote Job Entry process evolved into a method of sharing the mainframe's central processor among many users. Because of the central processor's high speeds, each user had the illusion that he or she had the sole use of that expensive machine. Such time-sharing required deft programming, and there was an overhead penalty involved in its use. But it was the only way to break out of the batch impasse until the mini-computer, and later the personal computer, came along. Pioneering work on time-sharing was done at several research institutions, especially at MIT and at Bolt Beranek and Newman (BBN) in Cambridge. Financial support came from the Defense Department's Advanced Research Projects Agency (ARPA), established in 1958 with its offices in the Pentagon. Around 1962 ARPA hired J. C. R. Licklider to head its Information Processing Techniques Office (IPTO), and Licklider became an enthusiastic supporter of the time-sharing projects then going on. Before moving to ARPA he had been at Harvard and at Bolt Beranek and Newman, and in effect he carried the "Cambridge view" of computing to Washington. As head of the ARPA office that dealt with information processing, he had the resources to push computing in that direction, which he did "without fanfare but with much determination."[12] For Licklider, time-sharing was not a means to bring mainframe computing power to customers; it was a way of building an "intergalactic computer network"—of creating a seamless net of interactive computers and people.[13] Eventually Licklider would succeed in creating such a network, called the ARPANET, the ancestor of today's Internet. Time-sharing was the spark.

Licklider's role in creating the ARPANET is well known, but we should note a less well-known story of how he inaugurated ARPA's support for time-sharing at MIT. In November 1962 he was aboard a passenger train taking attendees to Washington from an Air Force-sponsored conference in Hot Springs, Virginia. In the confines of the railroad cars "Lick" engaged in conversations with other participants, including Robert Fano of MIT. Other passengers on the train recall his moving from one seat to another, talking about his vision of networked computing to anyone who would listen. By the time the train arrived in Washington it was clear to Fano that an unsolicited proposal from MIT for time-sharing research would be funded. MIT prepared such a proposal within a few days, and ARPA accepted it. As Licklider's biographer says, "Nowadays, of course, this sort of involvement in a proposal would land Lick in jail for massive violations of federal procurement laws"[14] But that was typical of the way the IPTO worked, and no one argues with its results. One may keep this in mind when looking at the more recent scandals involving northern Virginia firms and their contracts with the Pentagon.

One final note about IPTO: a few years after its founding, it moved out of the Pentagon to a commercial building at 1400 Wilson Boulevard in Rosslyn, where it remained through the 1990s. Many have claimed "fatherhood" of the Internet, but we can say with assurance that we know *where* the Internet was conceived and funded: in a modest office building in Rosslyn. It stands within a few blocks of the former passenger terminus of the W&OD railroad, although that station was long gone by the time IPTO moved there.[15]

As Licklider was supporting an intergalactic computer network, the commercial services firms were adopting a more prosaic form of time-sharing. General Electric sold its time-sharing business to Honeywell in 1970, setting the stage for Honeywell's presence in Tysons Corner. IBM developed a time-sharing system for the System/360, and this became the preferred product for the service companies who were wedded to IBM products.[16] Among the local firms was U.S. Timesharing, founded by Mario Morino while he served as a U.S. Navy officer at the navy's headquarters next to the Pentagon. Morino later founded Morino Associates and become a major force in the computer services business along the Dulles Toll Road.[17] Companies that used the IBM system were less interested in the advanced MIT research on time-sharing, or in the ARPANET, as IBM had little to do with either.

The intense interest surrounding ARPANET's development obscures the importance of the day-to-day software houses, who used IBM mainframes to serve the vast federal bureaucracy. Still, ARPA's role in the creation of the Internet was substantial, and the Internet's creation is an advance in technology comparable to the development of electric power, the telephone, or the automobile. Those who suffered through the collapse of Internet stocks after 2000 may be skeptical, but the development of those other technologies went through similar phases. The hyperbole surrounding the Internet's creation can be stripped away, and one still finds an incredible accomplishment. But we must keep in mind that ARPANET's invention took place in the context of a more prosaic, but still significant, amount of more traditional computing and communication. The development of a technology is not a linear process. It does not proceed in a direct fashion from physical theory to lab prototype to production to market acceptance, ending with its social impact.[18] A linear model suggests a path from the theoretical studies at RAND to Licklider and ARPA's support for computer-based communication to the ARPANET, and then to the Internet. That leaves out too many other dimensions to the story. Janet Abbate, an historian of the Internet, makes a similar point: when the ARPANET was first built and had achieved a critical mass of nodes, few users knew what to do with it.[19] It had to be sold to skeptics.

At the insistence of Robert Kahn and Larry Roberts of IPTO, a network demonstration was staged at the First International Conference on Computer Communications, held at the Washington Hilton in October 1972. The network had been running in a rudimentary fashion for several years, and in 1972 it was still very much an experiment, but Roberts and Kahn felt that it needed a push to demonstrate to the computer science world that it was real, and that it worked.[20] By most accounts the demonstration was a success, but even afterward, the ARPANET did not suddenly gain acceptance. It was only with the invention of e-mail by Ray Tomlinson at BBN, also in 1972, that the network finally found a "smash hit," in Abbate's words. Tomlinson's program, which included his choice of the @ sign to distinguish between a user and its host computer, is acknowledged as one of the greatest computer "hacks" of all time (at that time the term "computer hacker" was still a compliment). It is also a cultural phenomenon, the implications of which we are only beginning to understand. IPTO managers did not foresee that e-mail would be a central part of their network, but they quickly recognized its value.

The creation of a networked community, facilitated by ARPANET, required the social input of its users, whose role was as important as that of the engineers, programmers, administrators, and, yes, politicians—including Al Gore, who provided government support.[21] But the users of the ARPANET were a restricted set. One had to be associated with a military-sponsored laboratory or an ARPA-supported university department. Many computer scientists from good universities were shut out of this network if they lacked such connections. As resistance to the Vietnam War escalated in the early 1970s, the easy informal access to DoD-funded programs on university campuses began to fray. The success of the 1972 network demonstration led some at BBN to form a subsidiary, called Telenet, to make packet-switching available to commercial customers. Larry Roberts left ARPA to become the head of Telenet, and by 1975 it was providing packet-switched services to seven U.S. cities.[22] Telenet's offices were in Reston, on Sunrise Valley Drive. The company expanded through the rest of the decade in spite of legal roadblocks placed in its path by the established telecommunications companies. It was sold at a profit to General Telephone and Electronics in the late 1980s.

THE PERSONAL COMPUTER AND PERSONAL NETWORKING

Another series of events, happening at the same time, also falls outside the linear model of Internet development. In 1975 an Albuquerque electronics hobby shop announced it had created a personal computer. The "Altair" was extremely limited in its capabilities, but it showed enough promise to lead a young Bill Gates to drop out of Harvard and move to Albuquerque, where he and Paul Allen cofounded their company Micro-Soft, later Microsoft. The Altair quickly faded, but in 1977 other personal computers with more capability were introduced. These too were of limited power but they did offer color graphics and game-playing capabilities—even the most privileged users of IBM mainframes did not enjoy that. They began selling in massive quantities to hobbyists and enthusiasts who wanted computer power but who had no institutional means of acquiring it. Not surprisingly, games were among the most popular uses although the trade press predicted that serious applications were not far behind. These users occupied an alternate world from those who had access to the ARPANET. Personal computer owners could use them as they saw fit, but it was a challenge to program these machines to perform rudimentary operations that the ARPANET community took for granted.

Into this turmoil stepped an entrepreneur named Bill von Meister. His biographers talk of a man with a flair for fast living whose mind was always full of ideas, racing ahead of any ability to put them into practice.[23] Among those ideas was a time-sharing service, but one aimed at the public, not business or professional customers. He was not the first to think of this—a service called CompuServe had been selling surplus time on mainframes to consumers via telephone lines since 1969. CompuServe's pricing was lower than comparable professional services like Lexis (intended for the legal profession and founded in 1973), but it was not low enough, nor was the access easy enough, to break into a mass market.

Based on an earlier, failed telecommunications venture, von Meister saw that advances in technology would open up a personal market and allow him to prosper. In 1978 he founded a service, The Source, that used the personal computer as a personal communications device. He set up offices in the Westgate office park at 1616 Anderson Road, just off Route 123.[24] The Source struggled financially, but it did attract the attention of computer visionaries, who saw in it a glimpse of a future that has since come to pass with the World Wide Web. Stewart Brand's 1984 edition of the *Whole Earth Software Catalog*, for example, devoted a lot of attention to it, arguing that this was the most important of all personal computer applications.[25] A later publication by the Whole Earth group said that The Source was to CompuServe as the Avis car rental company was to Hertz: "they try harder."[26] Neither Whole Earth publication, even the one published in the 1980s, made any mention of the ARPANET or Internet—another indication of the gulf between personal computers and mainframes.

The Source soldiered on, although von Meister was ousted in 1979. Compuserve acquired it in 1989 and folded The Source into its own service. But von Meister did not abandon his goal. His vision of networking was the counterpart of the vision Steve Jobs and Steve Wozniak had for personal computing. Their computer, the Apple II, overturned a industry based on charging for processor cycles, even if its power was so limited that people had a hard time figuring out what to do with one. Von Meister recognized that a similar shift was happening in telecommunications. That industry was dominated by AT&T, a regulated monopoly that slowly meted out technical innovation to its customers. In 1983 AT&T was broken up into separate local and long-distance companies, but even those who engineered that break-up did not see, as von Meister saw, the revolution that was about to transform telecommunications.[27] The Source is long forgotten. Most histories of the Internet ignore it, as The Source was not part of the linear sequence of events leading from ARPA's initial research to the Internet as it emerged in the 1990s. But the experience that today's Internet surfers enjoy stems more from von Meister's vision than from any other place.

After his ouster from The Source, von Meister thought of a scheme to deliver high-quality music to the home, using hardware of his design. This, too, was years ahead of its time, and failed. To help develop the system he enlisted the help of Marc Seriff, a programmer who had worked on the ARPANET. Meanwhile, von Meister kept coming up with other combinations of personal computing and telecommunications: if not remote delivery of music, then computer games to the home. In 1982 he founded a new company, Control Video Corporation, to market the latter. That showed promise, but in early 1983 the market for personal computer games collapsed, taking ControlVideo with it. Along the way von Meister enlisted the help of two individuals who would have more success in bringing his ideas to fruition: Steve Case, a young marketing executive at Pizza Hut, and Jim Kimsey, a Washington, D.C. entrepreneur who had made a modest fortune opening a series of downtown bars catering to the newly coined "Yuppies" flooding the city.[28] By late 1985 von Meister was shut out of the process, but Kimsey, Case, and Seriff carried on, founding yet another company called Quantum Computer Services, also located in Tysons Corner. Its focus was on the Commodore 64, an inexpensive computer popular among gamers. The service offered a network where members could chat, get news, play games—in short, a poor man's ARPANET, with content to match. Users accessed the network by dialing a local telephone number, which was in turn connected to Quantum's computers over privately leased telecommunications lines. The local call was free but users paid variable fees according to what services they accessed, the time connected, and so on. In contrast to the ARPANET, it was a private, centralized network.

Quantum looked at first like it was going to follow the other companies into oblivion, but Case and Kimsey pressed on. In 1989 they changed the company name to America Online (AOL) and opened its service up to other computers, including IBM and compatible PCs running DOS and Windows operating systems. At first AOL was a distant third in competition with the venerable CompuServe and with a new company called Prodigy, but by mid-1995 AOL was well ahead. By 1998 AOL had over 12 million subscribers and was by far the largest portal for laypersons to access computer networks.[29] Before 1996, subscribers paid by the minute, which led to a handsome revenue stream. As it expanded, AOL moved to a succession of larger offices, from a modest building on Aline Avenue to a suite at the end of Westwood Center Drive, both in Tysons Corner, to a huge sprawling campus north of Dulles Airport. (All of those locations are, coincidentally, close to the W&OD rail-trail.)

Among all the factors that led to AOL's success was its ability to instill a sense of community among its subscribers, something that people noticed about The Source. The key to that was AOL's chat service. J. C. R. Licklider had a vision of community in mind when he wrote of an "intergalactic computer network" years before. He did not foresee the advent of the personal computer, which could make that network far more widespread than he imagined. His successors at ARPA began building such a network, and in

FIGURE 7.2
AOL headquarters, Ashburn, circa. 2004. Photo by author.

the process overcame enormous technical hurdles, but the ARPANET was never widely accessible, not even to computer scientists on college campuses. By contrast, AOL chat rooms operated out of the limelight. Unpaid volunteers moderated them and gently shaped their content. Over time the content of those rooms degenerated, illustrating a modern version of Gresham's Law, but initially many of them were of surprisingly high quality. By signing up with AOL, 12 million people could sit at the Round Table of the Algonquin Hotel, or engage in a witty discussion at Mabel Dodge Luhan's salon. The service had many critics. Veterans of the early days of the personal computer could not stand how AOL opened up networking to hoi polloi, forgetting that the whole point of the personal computer was to make computing power widely available. Other criticized the often crude content of the chat rooms—but what did those who sat at the Round Table talk about? AOL developed other services besides chat that instilled a sense of community; taken together, these drove the service to a dominant position among computer networks.

FROM ARPANET TO INTERNET

ARPANET connected computers using software called a Network Control Program (NCP). It was the equivalent of the marks and stamps on a letter: name, address, ZIP code, postage, return address—each located in a defined place and with a well-defined form. These marks and their placement follow rules and are distinct from the contents of an envelope, which have no prescribed form. The networking equivalent was called a "protocol," following the original meaning of that word: a piece of parchment or vellum attached to a scroll to identify its contents. By 1983, as the network grew, the ARPANET switched to a more sophisticated protocol, the Transmission Control Protocol/Internetwork Protocol (TCP/IP). The change was painful and many sites resisted it, but ARPA's control over the network ensured that the transition would happen. The new protocols helped make the transition from the ARPANET to the Internet, which connects a wide variety of computers and other devices. TCP/IP is a major factor in the ability of the Internet to scale up to millions of nodes. It is still in use.

Many researchers had a role in developing and refining the concepts for TCP/IP, but among them two have received the most credit: Robert Kahn and Vinton Cerf.[30] We have already encountered Robert Kahn, who led the effort to demonstrate the ARPANET at the Washington Hilton in 1972. After that demonstration, Kahn joined IPTO as a program manager and soon found that a new challenge awaited him, that of extending the ARPANET to a much more open-ended Internet—a network of networks. That led to his co-authorship of the TCP/IP protocols. He stayed at IPTO through 1985, serving as director in the last six years of his tenure there. After leaving IPTO he founded the nonprofit Corporation for National Research Initiatives (CNRI), devoted mainly to issues of Internet governance and research. Some have called it a civilian ARPA: devoted to securing funding and other support for advanced technology, but privately funded. Its offices are in a modest but attractive building in a wooded setting on Preston White Drive in Reston. Like AOL, CNRI's offices are close to the W&OD bicycle path and can even be seen from the path in the fall and winter when the trees are bare.

Vinton Cerf received a BA in mathematics from Stanford in 1965 and was at UCLA between 1967 and 1972. He returned to Stanford, where in 1973 Kahn enlisted him to help define the new protocol.[31] He initially received a contract to do this work while at Stanford, then in 1976 he joined ARPA, serving as a program manager for the Information Processing Techniques Office.

In 1982 Cerf left ARPA to become a vice president at the telecommunications company MCI. That company was founded in 1963, initially to provide communications between Chicago and St. Louis. The "M" stood for "microwave": it used radio towers that were independent of the AT&T network. AT&T used its political clout to fight the proposal, and it was not until 1971 that MCI was able to set up its network.[32] The cost of building the network was trivial compared to the legal costs incurred fighting AT&T. In 1968 MCI's chief executive, William McGowan, set up offices in downtown Washington, D.C. According to legend, he chose an office close enough to the FCC headquarters so that MCI lawyers could walk there. MCI spent so much of its time fighting AT&T in the courts that the press dubbed the company "a law firm with an antenna on the roof." McGowan's persistence prevailed, and by 1980 it became the first company outside AT&T to offer long distance telephone service.[33]

While MCI's main offices were in downtown Washington, its back offices and engineering facilities were in Reston, off Reston Parkway. Thus, during most of the 1980s and 1990s, the two architects of TCP/IP worked at different offices in Reston not quite within sight of each other. Around 2000, MCI moved to a sprawling campus in Dulles (Ashburn). It remains there today, although it has had a turbulent history—it was WorldCom, filed for bankruptcy, and was then taken over by Verizon. This is an ongoing saga, although MCI will probably continue to have a presence in Ashburn.

In the decade from 1985 to 1995 the Internet was born. Unlike the ARPANET, the Internet is a network of disparate networks, not just a single network, and it is open to commercial and personal activities. The Internet also reflects the changing economics of computer technology. The initial plan was to connect expensive and scarce mainframes. Now the Internet mainly connects desktop computers that are linked to one another in small networks, say in an office or on a campus; it also now connects even smaller devices such as cell phones. These local networks are connected to larger networks spanning a region, and then to Internet "backbones": high speed fiber optic channels across the United States, with connections to the rest of the world.

FIGURE 7.3
MCI headquarters, Ashburn. *Insert:* The facility is located at the intersection of the
Loudoun County Parkway and UUNET Drive. Photos by author.

The transition to the Internet began in the late 1980s with the establishment of a backbone, administered by the National Science Foundation and
built under contracts by MCI and IBM.[34] ARPANET continued to operate,
although by 1988 it was obsolete; it was retired in 1990. Regional networks,
typically administered by university consortia, were connected to this backbone at a small number of access points, called Federal Interchange Connections (FIX) for government and university users, and Metropolitan Access
Exchanges (MAE) for commercial users. Commercial users were allowed
to connect, but they could not engage in commercial activity. That was

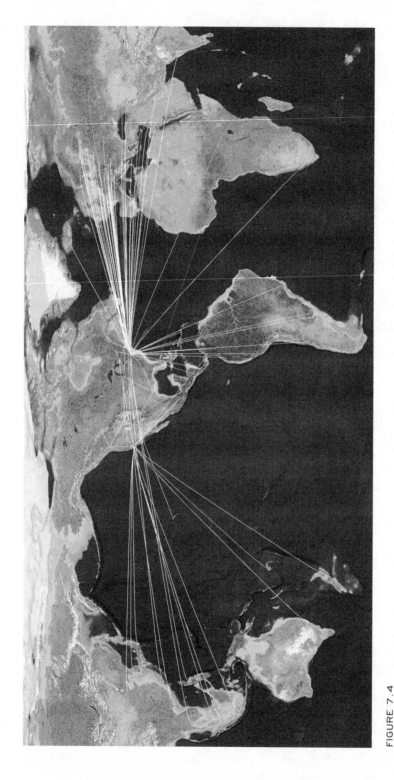

FIGURE 7.4

The National Science Foundation-sponsored NSFNET. Note the concentrations on the east and west coasts of the United States. Well into the 2000s, Internet traffic from one northern European country to another would pass under the Atlantic twice, after being switched at locations on the U.S. east coast. After the September 11, 2001 attacks, which damaged a switch in lower Manhattan and disrupted intra-European Internet traffic, backup switches were installed in Europe. *Source:* National Science Foundation.

prohibited by the "Acceptable Use Policy," a clause in the National Science Foundation's authorizing legislation. An amendment to that clause allowing such activity was passed in 1992, and when that became law, one could say the modern Internet age began. That amendment was sponsored by Congressman Rick Boucher of Virginia's Ninth Congressional District (in the southern part of the state). Senator Al Gore was involved in this activity: he sponsored a bill providing funds for a high-speed network to connect scientific researchers, mainly on university campuses. Even before that legislation passed, commercial services began to connect to the backbone—one of the first was MCI Mail, a commercial e-mail service inspired by Vint Cerf, who was at MCI at the time.[35] This and other services did not run afoul of the Acceptable Use Policy, as they purported to be engaged in research activity.

Other commercial providers joined MCI. Among them was Performance Systems, Incorporated (PSINet, pronounced "pee-ess-EYE-net"), founded in 1989 in New York by William L. Schrader. In 1991 it moved to Virginia and focused on providing commercial access to the Internet.[36] It grew rapidly, its stock a mirror of the Internet bubble of the late 1990s, rising to a dizzy height and then crashing back to ground. The company attracted local attention in 1999 when it bought the naming rights for the Baltimore Ravens football stadium. NFL fans were just getting used to pronouncing the name when PSINet filed for bankruptcy in 2001 (the stadium is currently named after a local bank). During its heyday, its offices were in a modest office park in Herndon, adjacent to the W&OD rail-trail.

Another important commercial provider was UUNet (pronounced "YOU-you net"), founded in 1987 by Rick Adams, who had been exposed to the ARPANET while working for the Defense Department and SAIC, both in Arlington. Once the Acceptable Use Policy was amended, UUNet moved into commercial networking under the aggressive leadership of John W. Sidgmore, who took over as president after having worked at General Electric's Information Services in Rockville. The name came from an obscure early networking program, UUCP, which connected computers that used the Unix operating system. Unix remains important among the machines that glue the Internet together, but most users seldom encounter it. In a complicated set of transactions beginning in 1995, UUNet merged with a fiber-optic company called Metropolitan Fiber Systems; the following year that combination was bought by the Mississippi telecom company WorldCom, which also bought MCI in 1998.[37] After the dust settled at the turn of the millennium, WorldCom was bankrupt and its CEO was under indictment. MCI emerged with its old name back, owning both UUNet and Metropolitan Fiber Systems.[38] UUNet's headquarters had been located in Fairfax for much of this period. It is now folded into MCI's campus in Ashburn on UUNet Drive, not far from AOL.

MAE-EAST

This convoluted story has been condensed because of the way Internet connections evolved after commercialization was allowed. To summarize, MCI ended up with ownership of the major portion of the National Science Foundation's backbone, to which connections were limited to a select few places called network access points. The most important of these was MAE-East, set up by Metropolitan Fiber Systems in 1992. It was located in the parking garage of an office building in Tysons Corner, a block from AOL's initial headquarters.[39] One of the defining features of packet-switched networks like the Internet is its dispersal of switching. But when the NSF built its network as a replacement for the ARPANET, it was more concerned with the research value of the network than with security implications. Thus for a while nearly all traffic moved through a few access points: FIX-East in College Park, Maryland; MAE-East in Tysons Corner; MAE-West in San Jose; and FIX-West in Mountain View, California.[40] Some Internet analysts estimated that in the 1990s MCI routed over 70 percent of domestic Internet traffic on its backbone, with over 50 percent of all traffic passing through MAE-East.[41] This arrangement did not last as other, parallel backbones were built by other telecom companies, and the MAEs were dispersed to other sites along the Dulles Toll Road and throughout Silicon Valley. The original facility in Tysons Corner may no longer be in use.

The location of MAE-East brings us back to the question that began this chapter: does geography matter? Another way of asking the question is to compare the construction of the Internet with other large technological systems built in the United States in the twentieth Century. One can visit the massive dams along the Columbia or Colorado Rivers in the West, or the facilities of the Tennessee Valley Authority in the South, and understand the magnitude of the engineering required. That is not true of the Internet, although it is an achievement of equal magnitude. The public is not allowed to visit MAE-East, and even if it were, there is not much to see. The original

MAE-East contained racks of servers and routers that were hardly different from what one might find in the IT room of an office building or a college campus. The fiber-optic cables carrying the traffic are thin and unimpressive. They are installed in plastic tubes and buried under highways and railroad rights-of-way (the latter favored because of their isolation—other utilities are unlikely to dig nearby). A lot of the Internet is in such conduits between Tysons Corner, Reston, Herndon, and MCI's facilities in Ashburn. Some of it is buried along the W&OD; some is under streets in Herndon, Routes 7, 123, and other local roads. The locations are not publicized, but fiber-optic cables carrying Internet traffic are everywhere these days.

Fiber-optic cables are not the same as silicon chips, but their capacity follows a rate of progress similar to what is known in the computer field as Moore's Law. That is an empirical observation that computer chip density doubles about every eighteen months, and has done so steadily since 1965. Fiber's bandwidth also increases at an exponential rate—sometimes faster than the market can absorb it, one cause of the telecom collapse of 2000.[42] The first transatlantic telephone cable, installed in the 1950s, carried thirty-six simultaneous telephone calls and cost over $30 million in 1950s dollars. A modern fiber-optic line is the equivalent of about 200,000 1950s-era transatlantic cables.[43] In 2005, a telephone company began offering fiber to homes in selected Washington suburbs; these have a bandwidth equivalent to 500 voice channels.[44] The physical structure of the Internet looks like an ordinary telephone wire, not like the Grand Coulee Dam. But looks are deceptive: building the Internet took just as long, if not longer, and cost as much, if not more, than those projects along Western rivers.[45]

THE DOMAIN NAME SERVERS

As the Internet took form in the mid-1980s, its creators came up with the now-familiar addressing scheme. This was a set of three-letter codes indicating the type of organization where the node was located: the familiar .com, .org, .edu, and so on. These codes were at the rightmost part of an address. Initially the educational (.edu) and military (.mil) nodes were most common, but by the mid-1990s the commercial (.com) node overwhelmed all others. This scheme was supplemented by a set of suffixes indicating the country where an Internet node was located, based on the International Organization for Standards' two-letter codes for countries (For example,

".fr" for France, or ".us" for the United States). The computers themselves translate these codes into a long binary number, which they use to route packets efficiently across the network. On the rare occasion a code is written out, it usually appears in "dotted–decimal" form, like so: 160.111.70.27.

The beauty of this system stems from its hierarchical structure. A computer tasked with routing a packet only has to decode the address until it has enough information to send it on its way, just as one need not dial a country code when making a local telephone call. Nevertheless the Internet needs a master directory to keep all the addresses current and accurate down to the lowest level. This concept resembles a telephone directory, which is published once a year, in local editions. People look up local numbers in the phone book, but few of us have directories for all the cities in the country, never mind the rest of the world. And by the time a book is printed, some numbers in it have changed, others have been disconnected, and new numbers have been added.

Initially the Defense Department kept this Internet master phone book equivalent at a site in Chantilly. In 1993 the National Science Foundation took over that job; by this time the .com suffix was eclipsing the others. The NSF in turn awarded a contract for the management of the top-level domains .com, .org, .net, and .edu to Network Solutions, Inc., a northern Virginia company founded in 1979.[46] The origins of this company are obscure, and I have not been able to find information on how it competed for that contract, one of the most critical in the history of the Internet.[47] Network Solutions now had the task of managing the Domain Name Registry—ensuring that each Internet address was unique and that messages to each address could be efficiently and reliably routed over all the networks comprising the Internet. The question of who had the authority to manage this aspect of the Internet is not easy to answer, but Network Solutions was guided at first by an entity called the Internet Network Information Center, or InterNIC, set up with AT&T. Beginning in September, 1995 Network Solutions began charging $50 a year to register a name. In 1998 its authority to manage domain names was transferred to the Internet Corporation for Assigned Names and Numbers (ICANN), administered by the Commerce Department. ICANN's authority was not formally established, however, and therefore in some sense one could argue that no one runs the Internet or has the authority to assign domain names.[48] Eventually .com names would be bought and sold for enormous sums of money, and corporations sued individuals who registered names that seemed to infringe on their intellectual property.[49]

In that context it seems incredible that at first the assignments were handled so informally—for years by just Jonathan Postel and his assistant at the Information Sciences Institute in Marina del Rey, California. Postel had no formal authority; he simply took on the job and did it—and he did it well. That kind of arrangement could not last forever, leading to the Network Solutions contract.[50]

InterNIC, Network Solutions, and AT&T set up thirteen computer systems, called "root servers," to hold this directory. In 1998, five of those thirteen were in the Dulles corridor: one at Boeing Court in Tysons Corner, two at Network Solutions' offices in Herndon, and one at PSINet's offices in the same Herndon office park. Although not specified, the other was probably at the InterNIC site in Chantilly. Of the remaining eight, two were in Maryland (College Park and Aberdeen), three were in California, one was in the UK, one was in Sweden, and the last was in Japan.[51] Thus of the thirteen root servers, seven were in the Washington region, and *three* in the same office park in Herndon.

One of the root servers was designated the top-level, or "A" root server: the "*capo di tutti capi.*" This held the "root zone file": the master directory listing. At least twice a day the file was sent to the other servers, which in turn sent the file around the world to lower-level servers, and so on. If this file were corrupted or the server physically damaged, any of the other root servers could quickly be designated as the top-level server with minimal disruption to the Internet. This has happened—at least once by a deliberate attack—but in each case the damage was contained.[52]

The A root server was one of the two installed at Network Solutions' offices in Herndon. The buildings in this park were all low and had few, if any, markings on them. Bicyclists used their parking lots to gain access to the W&OD rail-trail, which runs by one side. This office park was one of the places (along with MAE-East) in northern Virginia where cyberspace was connected to the ground.

In 1995 SAIC bought Network Solutions, and five years later it sold the company to the Silicon Valley firm Verisign. The timing was fortuitous: SAIC bought Network Solutions just as people were clamoring to buy up .com names, and it sold the company at the peak of the Internet bubble. Verisign moved the A root server from Herndon to an office building in Sterling.[53]

The W&OD rail-trail opened in sections beginning in 1974, when little of the Dulles Corridor was developed. The trail was completed to Leesburg by 1985, when the corridor began to grow rapidly. As Internet Alley grew, so did office parks and high-rises along the trail. Today the W&OD trail threads its way though a canyon of high-rises in downtown Reston. This kind of growth is a threat to the trail's future, but those who work in the technology companies along its route are among its most enthusiastic users. They band together to protect it. Near where the Bowman distillery stood in Reston, a sign proclaims the advantages of an extended-stay hotel: "Homestead Village: Running during lunch. Running your company during the day."[54] Although not touted as one of the trail's amenities, it is the best way for a visitor to see what little there is to see of the physical manifestation of the Internet—one of the great engineering projects of the twentieth century.

THE "PEACE DIVIDEND"

By the late 1990s AOL had millions of customers, each of whom was paying about $20 a month for service. The Tysons Corner contractors had one customer (the federal government), and their fortunes often depended on winning or losing a single contract, for which they would charge millions of dollars. On the surface these business models could not be more different, even if both types of company relied on computer networking and software development. During the 1990s a number of factors converged to heighten this contrast. Then, a series of events after the turn of the millennium reversed it.

The 1990s began with the end of the Cold War. With the election of Bill Clinton as president in 1992, there was talk of a "peace dividend": using money formerly spent fighting the Soviet Union for other, presumably more peaceful, purposes. Military budgets declined. Some of the Tysons Corner contractors had financial problems, but as Earle Williams noted, government contracting never went away. Those events coincided with a financial crisis in the savings and loan industry and a slowdown in the real estate market. Vacancy rates for northern Virginia office buildings went up. It became harder to make money in commercial real estate and the associated construction trades, though Earle Williams's comment about military contracting also held here: well-managed real estate developers continued to make money through this cycle.

While that was happening, employees who had received stock in companies like AOL or PSINet suddenly found themselves very wealthy, as any stock having to do with the Internet shot up to absurd valuations. The local and national press responded with stories about how the contractor economy of northern Virginia was dead; long live the Internet economy.[55] The enormous paper fortunes amassed by these people caused further troubles. In an earlier day, developers catered to the defense contractors' need for discretion. They built buildings that were comfortable but not ostentatious, as their tenants had to deal with people who worked in the Pentagon, a government building whose austerity mirrored its tenant's grim mission. The Internet companies had no such inhibitions. One developer of a building for a Internet company recalled putting in a luxury suite—including a shower—for an executive's dog.

On a more serious note, business leaders questioned whether northern Virginia had what it took to compete in an entrepreneurial climate. BDM and its brethren sold themselves as being more agile, more efficient, and more flexible than the government laboratories and civil servants they competed against. Now they were facing the same charge from the Internet companies. If a company has only one customer, it is not going to be strong in marketing, whether to consumers or to small businesses. Marketing has always been one of Silicon Valley's strengths, and Steve Case's marketing experience at Pizza Hut was key to AOL's success. Mario Morino, whose software company sold products to businesses as well as to government, noted this as well: for him the Washington region was hampered by a lack of marketing, executive, and managerial talent.[56] Earle Williams would not agree with that assessment, but he did acknowledge that BDM had trouble cultivating non-government sales. He noted, for example, that while in the commercial world it was common for a salesman to take a client and his family to a baseball game or to the ballet, the BDM sales force believed that such activities were forbidden when seeking a contract from a military procurement officer.[57] As early as 1986, well before the Internet bubble began, William McGowan of MCI noted a similar problem. To him, Washington entrepreneurs were afraid to take risks. In Silicon Valley nearly every veteran entrepreneur has been through one or more failed enterprises, but for McGowan, "Washington is populated with people more conscious of life's perils than its adventures."[58] Companies were springing up in Tysons Corner as he spoke, but he noted with scorn that people founded these companies *after* they had a government contract in their hands.[59]

However serious were the perils of military contracting during the Clinton presidency, they were offset by the wealth generated in Internet Alley. And local politicians and planners were happy to adapt to the change. Mario Morino called the region the "Potomac Knowledgeway," implying that Maryland was also a participant in the transformation. The Northern Virginia Technology Council coined the term "Techtopia" and began publishing a map showing the companies and their locations.[60] (Other terms were also coined, but for the reasons given earlier I prefer "Internet Alley.") When AOL bought the pioneering Silicon Valley firm Netscape in the fall of 1998, it symbolized the region's transformation.[61] Netscape's Web browser, released in 1994, had started the Internet boom. Now Marc Andreessen, Netscape's founder, multibillionaire, and boy wonder, was moving to Virginia and buying a house in McLean.

THE BUBBLE

On March 20, 2000 MicroStrategy, a Tysons Corner software company, restated the revenues it had claimed for the previous year. The over $12 million in profit that it had claimed suddenly became a loss of $30 to $40 million. The stock plummeted over the next few days, from a peak of $313 a share to about a dollar. On one day in March it lost 140 points, causing the value of stock held by Michael Saylor, the founder and CEO, to lose six billion dollars—a possible world record.[62] MicroStrategy was one of the first indications that the dot-com bubble was bursting. Ten days earlier the NASDAQ stock index closed at 5,060, a valuation it has not come close to since. Most of the deflation of that bubble happened in Silicon Valley, but northern Virginia had its share of troubled companies. Amazingly, MicroStrategy was not one of them; it survived. PSINet and WorldCom both declared bankruptcy, in 2001 and 2002. Teligent, a networking company with headquarters in the Tiffany building (Fairfax Square), also peaked in value in March 2000 at over $6 billion in market capitalization. It went bankrupt the next year. AOL had merged with the media company Time Warner in 2000; the combination survived, but its stock plummeted and Steve Case was forced to resign. After a few years the combination dropped "AOL" from its name. These were only the more visible signs of the bubble's bursting.

The local press stopped reporting how the Internet economy was replacing the tired old government-contract economy. Now the government contracts were seen as a shock absorber that would save the Washington region from total collapse. Nevertheless, the mood at places like the Northern Virginia Technology Council was anxious. Venture capitalists continued to meet and make deals at local clubs, although the number of new companies they funded was very small. The Tower Club remained a place to meet and greet people in the Internet and telecommunications fields, but the mood there was subdued. By a strange twist of fate, visitors to the Tower Club had to pass by the offices of MicroStrategy on their way to the top floor. That must have caused investors to take a hard look at any plans for start-ups being discussed at lunch.[63]

Still, business was being done. The Internet continued to grow in spite of the stock market collapse. Mario Morino used some of the money gained when he sold his software company to support a technology "incubator": a building in Reston where start-up companies could find modest office space, high-speed data connections, and a chance to put their ideas into practice.[64] Morino continued to have faith in the region's shift from defense contractors to an Internet-based economy.[65] A list of his tenants in April 2003 gives a flavor of the enterprise. Let the names speak for Morino's vision:

· Anuera
· Carlyle Venture Partners
· CyberMark
· e-know
· Engenia
· FBR Technology Venture Partners
· Grotech Capital Group
· mindSHIFT
· Netrepreneur.org
· Simplexity
· Updata
· VentureBank

The real estate market did not collapse. Retail shopping remained healthy, although the Tysons Galleria mall struggled and was sold to a Chicago-based company in 1992. Reston opened a "town center": an example of the New

Urbanist philosophy of building a shopping mall with storefronts opening onto a street, as if to recreate a downtown of the 1930s. The technology firms renting the upper floors of Fairfax Square, including Informix and Teligent, did not do well, but Tiffany, Gucci, and Hermes remained on the ground floor. In July 2000, the famous mail-order supplier L.L. Bean opened its first retail store outside of Maine at Tysons Corner Center.[66] Bean had built a reputation selling the Maine Hunting Shoe, wool shirts, and other simple fishing and hunting supplies. The décor of the Tysons store was more upscale and catered to shoppers who seldom got away from the suburbs, even if they drove four-wheel-drive sport utility vehicles. Bean's press release stated that the Tysons Corner store is closer to the Appalachian Trail than is its flagship store in Freeport, Maine. Locally the Appalachian Trail runs along the Blue Ridge, passing closest at Bluemont—coincidentally the terminus of the W&OD railroad. And in May 2001 Apple Computer announced that it, too, would open its first-ever retail store, also in Tysons Corner Center, not in Silicon Valley.[67] The role of the Internet and systems technology in the local economy was a factor in that choice. Both of these stores have been profitable.

The number of U.S. military troops deployed overseas dropped, but the United States continued to have a presence around the globe, and it continued to develop and deploy advanced weapons systems on the ground, in the air, on and under the sea, and in space. All those systems still required systems engineering and integration. George W. Bush's victory in the 2000 presidential election was welcomed by many defense contractors, who felt that the draw-down of military forces after the end of the Cold War was too drastic. Among the new president's actions was to renew an emphasis on ballistic missile defense, the activity that generated some of BDM's first contracts in El Paso forty years before. By this time BDM was gone, but its successors TRW and Northrop Grumman retained ballistic missile defense expertise.

The mood was cautiously optimistic in Tysons Corner on the clear, sunny morning of September 11, 2001. But from the upper floors of the office buildings, tenants could see the smoke rising from the Pentagon after it was hit by one of the hijacked airliners. Of the 184 people killed in that attack, 125 were in the Pentagon. Of them, 70 were civilians and 55 were members of the military services.[68] Four were employees of BTG and Booz-Allen & Hamilton, doing contract work for the Army. The collapse of the Soviet Union reduced the odds of an exchange of ballistic missiles delivering city-busting hydrogen bombs from one continent to another, but the events of that day were a reminder of Thomas Hobbes's observation, made in 1651, that "the [natural] condition of man . . . is a condition of war."[69]

The terrorist attacks led to the creation of a Department of Homeland Security. Its headquarters were in the District, but most of its contracts went to firms in Tysons Corner and Chantilly, adjacent to the National Reconnaissance Office headquarters. Maryland also benefited, as companies moved to Annapolis Junction, adjacent to the National Security Agency. The kind of services they provided was an interesting mix of the old and the new. As the press reported, the attacks might have been prevented had there been a better integration of intelligence agency knowledge before September 11. The Internet and telecommunications companies offered their services to remedy this deficiency. A hybrid emerged, one combining the classic OR expertise of the early Tysons Corner firms with the IT capabilities of the Internet companies. The combination fit well with an idea Pentagon planners had been talking about for a while: namely, the development of what they called "network-centric warfare."

The result was that traditional military contractors, now consolidated into a few giant aerospace companies, received a lot of business, although these contracts had a distinctive IT twist to them. In the 1980s, the coin of the realm was a retired military officer with some technical background who also knew the details of Pentagon procurement. In the 1990s the demand was for people who knew the Unix operating system and the C programming language. After 2001, the demand has been for people with both those skills, plus a security clearance. A tall order, hard to fill. Those companies who can supply such people are doing very well. The companies currently flourishing along Internet Alley are the ones who have been able to find and hire such people.

Let us return to Aesop's fable about the fly sitting on a chariot who exclaimed to the mule over what a mighty cloud of dust "they" were kicking up. Who was the driver who transformed Tysons Corner and the Dulles Corridor? Was it a group of hard-working individuals, or was it simply Uncle Sam?

It was both. Northern Virginia today manifests the vision and work of a number of remarkable individuals who shaped this region. They included real estate developers, local politicians, business executives, urban planners, and members of Fairfax County's administrative staff. Three of them deserve special recognition: Earle Williams, Gerald Halpin, and Til Hazel. Following them closely were George Johnson and John Herrity, followed in turn by about a half-dozen others. Also critical was the work of business groups, some ad hoc and others more formal: the Northern Virginia Technology Council, the Metropolitan Washington Airports Authority, and the Fairfax County Economic Development Authority, to name a few. George Johnson points out that the influence of these groups often ebbed and flowed, depending on the particular issue of the moment. Whenever it was necessary to marshal support around an issue, a group always seemed to be there. A few actions by these groups have taken on legendary status—the Noman Cole Report of 1976 to the County Board of Supervisors is the most famous.

Other factors one attributes to chance, but there always seemed to be individuals who acted to take advantage of those chance happenings. The rejection of Burke in favor of Chantilly as a site for a new airport was a chance event, but one that General Quesada seized on to ensure that Dulles Airport would be large enough to handle the traffic that he knew would come some day. The alignment of the Beltway near the country crossroads of Tysons Corner was not planned, but Gerald Halpin recognized immediately that it conferred advantages that other intersections lacked. One could not say it was a planned decision to locate the War Department in Virginia—that was driven by the urgency of a coming World War. As General Quesada's vision shaped Dulles Airport, so too did the vision of General Brehon Somervell and Colonel Leslie Groves—and FDR, who approved the final site and plan—shape not only the Pentagon building, literally, but also its place in the local economy.

Still, the mule pulling this chariot is the federal government.[1] It is a federal government that chooses to do much of its business by contracting out services to the private sector. That was an innovation in government and management, one engineered by Vannevar Bush and others. Histories of the federal government during World War II examine in detail the technical innovations that government supported, including radar, the electronic computer, the proximity fuze, and the atomic bomb. The innovation of contracting for services was just as significant.

Let us also return to the hypothetical visit by a delegation seeking to recreate northern Virginia's prosperity in its home region. The delegation will find the factors mentioned in Chapter 1: a well-educated workforce, good schools, a favorable natural setting, and local support for business; but to that list it must add that the region be close to the Capital of the Free World, near the headquarters of the most powerful military establishment in history. How does one recreate that?

A comparison to ancient Rome or imperial London is not far off the mark. At the end of the Cold War the United States stood alone as a world superpower. Other countries had advanced weapons and intelligence systems, but none came close to the combination possessed by the United States after 1990. Understanding northern Virginia in the twenty first century begins with the following partial list of technologies employed by the United States national security establishment:

· Land-based intercontinental ballistic missiles, tipped with nuclear weapons
· Nuclear-powered submarines, some carrying ballistic missiles
· Long-range bombers, based in the U.S. and at U.S-controlled facilities around the world
· Stealth aircraft
· Unmanned Aerial Vehicles (UAVs), cruise missiles, and precision-guided smart bombs
· The Global Positioning System (GPS), open to civilian use worldwide, but under U.S. Air Force control

· A network of reconnaissance satellites whose resolution is classified but is known to surpass what is available to civilians

· A network of signals intelligence (SIGINT), electronic intelligence (ELINT), and early-warning satellites encircling the globe

· The Internet, which is free and open to anyone like GPS, but is also under U.S. control

· A network of communications satellites, for example, MILSTAR and Iridium (the latter developed for civilian use but currently used by the military)

· A rudimentary ballistic missile defense system, not yet operational but being tested.

This is only a partial listing. The European Union, China, Russia, and a few other countries may possesses one or a few of these systems, but none comes close to the combination held by the United States. Even the combined military systems of all the nations just mentioned probably fall short of the U.S. arsenal.

Many of the items listed are concerned with information processing: command, control, intelligence, high-performance computing, encryption and decryption, image analysis. This is the modern counterpart of the roads that led to Rome two millennia ago. Those roads were built to carry information as much as soldiers or goods. A better comparison is to the British Empire at the beginning of the twentieth century. The British counterparts to the modern combination of aircraft, satellites, and fiber-optic cables were the Peninsular and Oriental Steam Navigation Company (P&O), the submarine telegraph cable, and radio stations. The first World War demonstrated the value of long-distance communications networks, and in the mid-1920s the British spent enormous sums of money establishing radio and undersea cable connections to its empire. The British cables, nearly all of which terminated in London, were routed through its colonies or territory it controlled, in contrast to German cables, which had intermediate switches at islands or countries that were neutral or even hostile after 1914.

The British radio network used so-called longwave frequencies—low frequencies that curved around the Earth, and were thus able to reach distant colonies. The technique required high-powered, large, and expensive transmitters.[2] In the 1920s radio amateurs pioneered an alternative technique, using shortwave frequencies to span long distances at a fraction of the cost. That made the longwave transmitters obsolete and threatened the economic viability of the undersea cables. Nevertheless, the British kept the cables in operation for national security reasons—cable traffic was, and remains, inherently more secure than radio signals sent through the ether. With the support of the government, a company called Cable and Wireless was founded in 1929; it operated undersea cables at a financial loss, but the Crown did not allow it to abandon them.[3]

These cables, though uneconomical, served the British well during the second World War. As in the 1914–1918 conflict, the Germans were at a disadvantage, even more so than they realized at the time. Not long after the outbreak of the war, Axis undersea cables were cut or otherwise rendered unusable. The Germans had to rely more on radio, which by its nature can be intercepted. The Allies were able to decrypt radio messages the Germans had encrypted with their "Enigma" and other coding machines. Some of this traffic was to U-boats and ships and could have gone no other way, but other encrypted messages were sent to Germans on land because they had no cable alternative. That these codes had been broken was something the Germans did not realize until long after they surrendered.[4]

The British example is relevant to the nexus of military and IT contractors in northern Virginia for several reasons. The superior British telecommunications facilities were crucial to the Allied victory in World War II, but they also allowed the Americans, who had less of an investment in obsolete technology, to surpass Britain in commercial telecommunications after 1945. Given the open protocols that make up the Internet, other countries may be able to exploit this computer network—invented in America and managed in the Virginia suburbs of Washington—against the interests of the United States. The British left a legacy of the English language and British respect for the law as its empire faded. Could the same happen to the United States, leaving others to exploit the openness of and universal access to Internet services?

The preceding chapter mentioned how visitors to the Tower Club had to pass by the offices of the company MicroStrategy—an inadvertent reminder of the perils of investing in the Internet. From 1999 until about 2003, motorists along the Dulles Toll Road were greeted by an equally startling sight—startling at least to an historian. That was the name "Cable and Wireless" on the top of an office building beside the road. The venerable British company was still in business: its "cable" was now made of glass fiber; its "wireless" was now relayed by satellite. It came to the Dulles Corridor when MCI was bought by WorldCom. As a part of that deal, Cable and Wireless purchased some of MCI's backbone service, and for a while it was a major supplier of Internet services in the United States.[5] A few years later, after the telecom bubble burst, the company fell on hard times and it divested this service. The ebb and flow of Cable and Wireless is tangential to our story. Yet its presence here serves as a useful reminder of the parallel with the British Empire, and with London, Rome, Paris, and other imperial capitals. The words "empire" and "imperial" are out of fashion, but the United States became the world's sole superpower after the collapse of the Soviet Union. It maintains that position thanks to a world-circling network of telecommunications, computers, command and control, and intelligence. It is practical to have this network managed near the seat of government.

From this perspective, we can better understand the relationship between the earlier, defense-oriented firms that moved to Tysons Corner after Sputnik, and the later, venture-capital–funded Internet companies that line the Dulles Toll Road. Both types of firms are involved in information processing activities that are crucial to the United States's position in the world. Much of the work is classified for military customers, but the commercial telecommunications and computing are no less important.

The U.S. military involvement in Iraq since 2003 has been controversial for many reasons. One point of controversy is the unprecedented use of private contractors to do work that soldiers did in previous wars—even in Desert Storm only a decade earlier. CACI and DynCorp are among the major contractors based in Virginia; many other local firms have contracts as well. There is no question that federal outlays to contractors have surged since 2001, and that a lot of that money has flowed into northern Virginia. One account gives an estimate of federal contracting increasing from $4 billion to $29 billion in the Washington area, then "in the past five years, that number has jumped to $52 billion, thanks to increased military and homeland-security spending following the terrorist attacks of September 11, 2001."[6] The size of the federal workforce has not increased over that time. Where does that money go? Not to military hardware, which is produced elsewhere. The *Washingtonian* magazine, which is concerned mainly with upscale shopping, real estate, and private schools, recognizes that it goes to the employees—mainly upper management—of the defense contractors. That magazine's photo spread of the mansions some high-level contractors have built for themselves is evidence of that.[7]

We may now sum up how this situation arose: northern Virginia benefited from a particular type of federal contract, which originated during the second World War as operations research, later evolving into systems engineering. OR was central to the deployment of complex weapons systems for all three branches of the military. This activity was centered in southern California. Many of the Tysons Corner firms were either founded by RAND Corporation alumni or had strong connections to it. So strong was the pull of the federal government that this activity took root in northern Virginia despite the lack of a world-class research university nearby. It did not develop around College Park, Maryland, where the region's top electrical engineering, physics, and computer science departments were located. If there was a critical time when this shift to Virginia began, it was between the orbiting of Sputnik in 1957 and the ascendancy of Secretary of Defense Robert McNamara in 1961. McNamara elevated systems analysis to the highest levels of defense planning. His critics—and there were many—found that the only effective way to counter McNamara's analysis of a weapon system was to develop a systems analysis of one's own. Regardless of how one feels about his approach, the result of McNamara's tenure was a surge in demand for this kind of expertise, a demand that has never slackened regardless of who has held the presidency.

Vannevar Bush created what his biographer called "federalism by contract," but Bush saw government work transferred more to universities than to corporations.[8] During the New Deal it was common for the federal government to take an active role in the planning and execution of large-scale technological systems. The most famous of those was the Tennessee Valley Authority.[9] Predating the New Deal were other federal laboratories doing similar work under federal control, including the National Bureau of Standards and the Naval Research Laboratory north of the Potomac River. Also in Maryland, the federally planned community of Greenbelt stands proudly as an alternative to the chaos of Tysons Corner. That model of active federal involvement in science and technology—as well as in land-use—did not prevail.

The only major exception was the National Aeronautics and Space Administration, founded in 1958 in response to Sputnik. By the 1990s, however, even NASA found itself contracting out many of its activities, including the launching of the Space Shuttle. Thus NASA made headlines in 2005 when its new administrator, Michael Griffin, announced that to send astronauts back to the moon and on to Mars, he wanted the systems engineering to be done in-house, not contracted out. NASA employees welcomed Griffin's statement, but the Professional Services Council, the trade association representing local for-profit systems firms, was skeptical that he could make that happen.[10] If NASA succeeds, could others follow? That would be a major shift, which might have an impact on the prevalence of high-paying engineering jobs in northern Virginia as compared to Maryland.

Montgomery County, Maryland also prospered from these contracts. Unlike Fairfax County, the lower portion of Montgomery County was already a prosperous suburb when the Beltway was completed. The county nevertheless has a share of federal contracts, many connected to the National Institutes of Health in Bethesda. But not all contracts have been in health care and biotechnology. Lockheed Martin's corporate headquarters is in Bethesda, a legacy of the original Glenn Martin plant's location outside Baltimore. IBM's Federal Systems Division was also located in Bethesda, and an IBM-founded plant in Gaithersburg is one of the county's biggest employers. Also in Gaithersburg was an electronics and space division of Hughes Aircraft, while a division of Fairchild Aviation and the laboratories of the Communications Satellite Corporation (Comsat) were farther out. These facilities have recently been closed down or absorbed by other companies.

Nevertheless, Montgomery County's economy lacks Fairfax's dynamism. Nowhere in the county, not even in the biotech corridor outside Rockville, does one feel the energy of Tysons Corner. Many residents like it that way, but according to Gerald Connolly, the current chairman of the Fairfax County Board of Supervisors, there is a severe new job creation imbalance between Virginia and Maryland. Connolly States that for every new job created north of the Potomac, *three* are created in Virginia, most of them in Fairfax and Loudoun Counties.[11] This ratio has increased, not decreased, since the bursting of the Internet bubble in 2000.

Fairfax and Montgomery Counties share many characteristics. Fairfax County has come of age and taken on the urban qualities that its Maryland neighbor had for some time. Both have run out of developable land. Neither county has any more dairy farms that one can buy and turn into office parks or residences. New development must either go farther west, to the next counties, or up: replacing older one- or two-story buildings with high-rise office towers and condominiums.[12] Politically the two counties are becoming similar. Both have a significant percentage of foreign-born residents, whose children are educated by the public school system. In the 2004 presidential election, Fairfax County voters preferred the Democrat, John Kerry, to the Republican, George W. Bush. Montgomery County had been consistently Democratic for years, but Fairfax County had not voted for a Democrat for president since 1964 (Bush carried the Commonwealth of Virginia by a solid majority, however).[13] For both counties, as for the region as a whole, the federal government still drives the economy.[14]

TRANSPORTATION

Those who shaped this region, when interviewed from today's vantage point, all agree that if there was one place where they failed to implement their vision, it was in developing a transportation network that kept up with the economic growth of Tysons Corner and the Dulles Corridor. In the 1970s one frequently heard people say, "If we can put a man on the moon, we can solve X," where X mainly stood for the problems of America's cities. The RAND Corporation and others tried to solve those problems, but their systems analysis only went so far. RAND and others failed because the politics were too refractory.[15] Designing and implementing new transportation systems also involves politics, but less so than the other problems RAND attacked. Systems engineers failed in the 1970s, but they need to try again. Tysons Corner is home to one of the world's greatest concentrations of systems engineering talent. Surely its inhabitants can do something about traffic.

The solution favored by Til Hazel is to build more limited-access highways, completing the plan for multiple beltways envisioned in the 1950s. One of those would extend the Fairfax County Parkway to the north and cross the Potomac into Maryland. There it would provide direct access to the biotechnology firms in Rockville and Gaithersburg, and connect with a planned outer beltway to Interstate 95 at Laurel. At present there is no bridge between the existing Beltway and Route 15, north of Leesburg—a distance of over 30 miles.[16] This plan would partially address the problem of Maryland residents commuting to the Dulles Corridor: a situation that leads to heavy congestion every day at the Beltway Potomac crossing. The plan has its critics, mainly among Marylanders living near the bridge's proposed location.

Superhighways are an old technology, little changed from the system that President Eisenhower advocated in his first term of office. Several advances in technology may help. One proposal is to charge tolls that are adjusted on a continuous basis depending on the congestion. Drivers would pay more money at certain times, but at least they would arrive at their destination at a reasonable time. A more radical proposal is to fit automobiles with radar and computers allowing them to travel very close to one another, each car inches from another's bumper. These "platoons" would give highways the traffic densities of a mass transit rail system while allowing people to drive their own vehicles. Differential toll collection is being studied and may be implemented soon, while platooning has yet to leave the drawing board.

After years of planning, in 2004 the region took a step toward bringing the Metro to Dulles Airport. When completed, it will finally make use of the right-of-way set aside by General Quesada forty-five years ago. But even with the right-of-way already in hand—one of the major expenses of any mass transit system—the Metro extension will be expensive. Initial estimates were in the range of $3.3 billion from West Falls Church to Dulles. The project has since been broken into two parts because of the cost, but more recently the cost estimates of the first segment, from Falls Church through Tysons Corner to the edge of Reston, have gone up. As steep as those costs are, they are not out of line with current Beltway improvement projects. Rebuilding the southern Potomac River drawbridge and the interchange with I-95 to the south have a similar combined cost—around $3 billion in current dollars.

The Washington Metro is clean, safe, and comfortable, but it too is an old technology. For Tysons Corner it has two major drawbacks. The first has been mentioned: the cost. Transit advocates reply that in place of the expensive heavy-rail Metro, they could build a cheaper light rail system—a return to the trolleys that once blanketed the region—or a modern bus system that duplicates many of the features of rail. Modern trolley cars and buses are more advanced than their 1920s ancestors, and light rail systems have been built for about one-third to one-half the cost per mile of a system like the Metro. But many drawbacks of the old trolleys remain, especially their tendency to get stuck in traffic along with automobiles when they run along or cross streets at grade. It will never again be practical to run trolley cars on the W&OD rail-trail, as tempting as that may be.

The second drawback is that rail transit is fundamentally ill-suited for suburban densities. A station at, say, Fairfax Square would be impractical for someone going to the Tysons Galleria, across Route 7. The distance is slightly beyond what people will walk, and there is no safe way to cross the streets. Having a subway station every few blocks works for midtown Manhattan, but not in Tysons. The planned Metro from Dulles airport to

FIGURE 8.1

A Danish proposal for an "RUF" (Rapid Urban Flexible) dual-mode personal automobile, circa 2005. It can operate as a personal automobile, then climb onto a monorail track and run in a train, with the high densities of fixed rail transit. Can this solve
the region's transportation crisis? *Source*: RUF International.

downtown Washington may have as many as twenty intermediate stops,
which will make the journey uncomfortable and slow for travelers going the
full distance.[17] The planned 2012 extension may have as many as four stops
in Tysons Corner alone; even with them, major sections of Tysons Corner
will not be accessible from a Metro station.

One answer may be a return to the Personal Rapid Transit (PRT) that was
aborted in Morgantown, West Virginia. That system has the feature of bypassing intermediate stops for those passengers going to a farther destination. Its
advocates compare it to a fleet of robot taxicabs, only they do not get stuck in
traffic. Recall the involvement of aerospace companies in Morgantown—
Boeing, the Aerospace Corporation, and the Jet Propulsion Laboratory—and
how that experience soured aerospace engineers from ever trying such a system again. It may be time to bring it back. Personal Rapid Transit seems never
to get beyond the prototype stage.[18] The Morgantown system has operated
with high reliability and in complete safety since it opened thirty years ago,
but no other PRT systems are in operation. Raytheon had a contract to build
a system near Chicago's O'Hare Airport in the 1990s, in a neighborhood similar to Tysons Corner, but the system was cancelled at the prototype stage.
ARAMIS, a system supported by the French government, was also cancelled

FIGURE 8.2

Arlington & Fairfax Railway "Auto-Railer," circa 1939: an early "Dual-Mode" transport system. The vehicle could run either on the rails of the Arlington & Fairfax transit system, or on the road. Its introduction did not stop the rapid demise of rail transit in the Virginia suburbs, however. *Source*: Alexandria Library, Special Collections, Ames Williams Collection.

at the same time.[19] This debate over PRT should not divert planners from an inescapable fact; namely, that the region faces a serious transportation crisis and it will take an imaginative solution to address it. If such a system is planned for the region, local history buffs will be quick to point out that Arlington tried, but abandoned, a combination bus and rail car, the "Auto-Railer," in the 1930s in a desperate attempt to keep rail transit competitive with the automobile.[20]

THE FUTURE

Tysons Corner is doomed. That is the opinion of urban planners who have analyzed Joel Garreau's study of Tysons Corner as an edge city. They argue that the New Urbanist movement, already a success in Reston Town Center, will spread toward office parks. Those wishing for a higher, pedestrian-friendly density will move away from Tysons toward Arlington, where CACI, the National Science Foundation, and DARPA already have their

FIGURE 8.3

Reston Town Center: the New Urbanism. Reston has adapted well to the changing landscape. Photo by Diane Wendt.

headquarters, or even to the District of Columbia, especially to a new federal and defense complex being built along the Anacostia waterfront. In 2004 planners began looking at transforming Tysons Corner into a more pedestrian-friendly, urban space like Arlington, but it will prove difficult to retrofit an urban grid onto what was built as an alternative to urbanism.

On the other hand, those who reject urban densities in favor of traditional suburbs will move the other direction, continuing a twenty-year trend of sprawl and "edgeless" cities. The concentration of Homeland Security contractors in Chantilly supports this trend.

A visitor driving through Tysons Corner today may conclude that Tysons Corner is not doomed after all. Developers are as busy as ever with the construction of new office buildings. An expansion of the Tysons Corner Center shopping mall is also underway. A casual survey the new office towers suggests a shift of tenants, from those engaged in engineering or defense contracting toward those engaged in banking and finance. That trend is a local equivalent of what happened in lower Manhattan in the past fifty years, where financial services companies have gravitated in spite of even greater densities and congestion.

The U.S. Constitution requires that the seat of government be located in the District of Columbia. The Defense Department is just across the Potomac but within the boundaries of the original ten-mile square designated for the federal city. If allowed to, lawmakers, defense officials, and even presidents would probably move to comfortable and secure suburban office parks, as the executives at MCI have done. But unless the Constitution is amended, Tysons Corner will continue to enjoy its advantage of being close to the seat of government while allowing its workers to live in an affluent suburb. The only serious threat to its existence would be if the federal government were to drastically curtail its defense and national security spending—not a likely prospect. Thus one can conclude that for Tysons Corner and its offspring the Dulles Corridor, the future is bright, whatever challenges it may face.

APPENDIX: LIST OF ACRONYMS

AJCC	Alternate Joint Communications Center
AMS	American Management Systems
ANSER	Analytical Services, Inc.
AOL	America Online
APL	Applied Physics Laboratory [Johns Hopkins University]
ARC	Atlantic Research Corporation
ARPA	Advanced Research Projects Agency (see also DARPA)
BBN	Bolt Beranek and Newman
BDM	Braddock, Dunn, and McDonald
BMD	Ballistic Missile Defense (see also SDI)
BoB	Bureau of the Budget (see also OMB)
BRAC	Base Realignment and Closure
BWI	Baltimore Washington International [Airport]
CAA	Civil Aviation Authority
CACI	California Analysis Center, Incorporated
C^3I	Command, Control, Communications, Intelligence. Variants: C^4I (add "Computers"), C^4I^2 (add "Information")
CBD	Commerce Business Daily
C-E-I-R	Council for Economic and Industrial Research
CDC	Control Data Corporation
CIA	Central Intelligence Agency
CICA	Competition in Contracting Act
CIT	Center for Innovative Technology

CIX	Commercial Internet [or Interchange] Exchange
CMS	Cambridge [Massachusetts] Monitoring System (see also TSO)
CNA	Center for Naval Analyses
CNRI	Corporation for National Research Initiatives
COBOL	Common Business Oriented Language
CSC	Computer Sciences Corporation
DARPA	Defense Advanced Research Projects Agency (see also ARPA)
DDR&E	Director of Defense Research and Engineering
DEC	Digital Equipment Corporation
DoD	Department of Defense
ERA	Engineering Research Associates
FAA	Federal Aviation Administration
FAR	Federal Acquisition Regulation
FCC	Federal Communications Commission
FFRDC	Federally Funded Research and Development Center
FIX	Federal Internet [or Interchange] Exchange
FTE	Full-Time Equivalent
GMU	George Mason University
GOCO	Government-Owned, Contractor- [or Corporate-] Operated
GPS	Global Positioning System
GPSS	General Purpose System Simulator
HOV	High Occupancy Vehicle
ICANN	Internet Corporation for Assigned Names and Numbers
ICBM	Intercontinental Ballistic Missile
IDA	Institute for Defense Analyses
IGY	International Geophysical Year, 1957–1958
INTERNIC	Internet Network Information Center
IPO	Initial Public Offering
ISI	Information Sciences Institute, Marina Del Rey, California
IT	Information Technology

JRDB	Joint Research and Development Board (see also RDB)
MAC	Machine-Aided Cognition, or Man and Computer
MAE	Metropolitan Access Exchange
MCI	Microwave Communications, Incorporated
MFS	Metropolitan Fiber Systems
MIT	Massachusetts Institute of Technology
MITRE	Massachusetts Institute of Technology Research Engineers
MLS	Multiple Listing Service
MX	Peacekeeper Intercontinental Ballistic Missile
NACA	National Advisory Committee for Aeronautics
NAICS	North American Industrial Classification System (replaced SIC)
NASA	National Aeronautics and Space Administration
NBS	National Bureau of Standards (see also NIST)
NCP	Network Control Program
NDRC	National Defense Research Committee
NIC	Network Information Center (see also INTERNIC)
NIH	National Institutes of Health
NIST	National Institute of Standards and Technology (formerly NBS)
NOAA	National Oceanic and Atmospheric Administration
NRO	National Reconaissance Office
NSA	National Security Agency
OMB	Office of Management and Budget (formerly BoB)
OR	Operations Research (in the UK, Operational Research)
ORO	Operations Research Organization
OSRD	Office of Scientific Research and Development
PLUS	Planning and Land Use System
PRC	Planning Research Corporation
PRT	Personal Rapid Transit
PSAC	President's Science Advisory Committee
PSINET	Performance Systems, Incorporated Network
RAC	Research Analysis Corporation

RAND	Research And Development
RDB	Research and Development Board (see also JRDB)
ROTC	Reserve Officers' Training Corps
SAIC	Science Applications International Corporation
SBA	[U.S.] Small Business Administration
SCI	Strategic Computing Initiative
SCIF	Sensitive Compartmentalized Information Facility
SDC	System Development Corporation
SDI	Strategic Defense Initiative (see also BMD)
SEAC	[National Bureau of] Standards Eastern Automatic Computer
SETA	Systems Engineering and Technical Assistance
SIC	Standard Industrial Classification (replaced by NAICS)
SWAC	[National Bureau of] Standards Western Automatic Computer
TCP/IP	Transmission Control Protocol/Internetwork Protocol
TRW	Thompson-Ramo-Woldridge
TSO	Time Sharing Option
UAV	Unmanned (or Unpiloted) Aerial Vehicle
UNIVAC	Universal Automatic Computer
USGS	United States Geological Survey
UUCP	Unix-to-Unix Copy Program
W&OD	Washington & Old Dominion [railroad, later rail-trail]
WMCCS	("Wimmicks") World Wide Military Command and Control System
WSEG	Weapons Systems Evaluation Group

SELECTED BIBLIOGRAPHY

Abbate, Janet. 1999. *Inventing the Internet*. Cambridge, Mass.: MIT Press.

Air Ministry [UK]. 1963. *The Origins and Development of Operational Research in the Royal Air Force*. Air Publication 3368. London: Her Majesty's Stationery Office.

Baucom, Donald R. 1992. *The Origins of SDI, 1944–1983*. Lawrence: University of Kansas Press.

Baum, Claude. 1981. *The System Builders: The Story of SDC*. Santa Monica, CA: System Development Corporation.

Baxter, James Phinney III. 1946. *Scientists against Time*. New York: Little Brown.

Bercovici, Liza. 1978. "The Study Merchants: Tysons Corner Offices Are Headquarters for Growing Army of Consulting Firms." *Washington Post*, November 12. C1, C6.

Brinkley, David. 1988. *Washington Goes to War*. New York: Ballantine Books.

Briody, Dan. 2003. *The Iron Triangle: Inside the Secret World of the Carlyle Group*. New York: John Wiley.

Brody, Herb. 1987. "Star Wars: Where the Money's Going." *High Technology Business*, December, 22–29.

Cho, David. 2002. "N.Va. Exhibit Examines History at a Crossroads." *Washington Post*, May 19.

Collins, Martin. 2002. *Cold War Laboratory: RAND, the Air Force, and the American State, 1945–1950*. Washington, D.C.: Smithsonian Institution Press.

Commission on the Future of the United States Aerospace Industry. 2003. *Final Report.* Arlington, VA.

Computerworld. 1996. "Computerworld's Top 25 Integrators by Revenue." February 26, SI/19.

Council on Economic Priorities. *Star Wars: the Economic Fallout.* Cambridge, MA: Ballinger. 1987.

Day, Kathleen. 1994. "Riding Herd on the Bad Guy Image of 'Beltway Bandits'." *Washington Post,* February 9.

Deiter, Ronald H. 1985. *The Story of Metro: Transportation and Politics in the Nation's Capital.* Glendale, CA: Interurban Press.

DeVorkin, David. *Science With a Vengeance: How the Military Created the U.S. Space Sciences After World War II.* 1992. New York: Springer.

Dickson, Paul. 1971. *Think Tanks.* New York: Ballantine Books.

Dyer, Davis. 1998. *TRW: Pioneering Technology and Innovation Since 1900.* Boston: Harvard Business School.

Galison, Peter and Bruce Hevly, ed. 1992. *Big Science: The Growth of Large-Scale Research.* Stanford: Stanford University Press.

Garreau, Joel. 1991. *Edge City: Life on the New Frontier.* New York: Doubleday.

Goldberg, Alfred. 1992. *The Pentagon: The First Fifty Years.* Washington, D.C.: Historical Office, Office of the Secretary of Defense.

Grava, Sigurd. 2003. *Urban Transportation Systems: Choices for Communities.* New York: McGraw Hill.

Green, Constance McLaughlin. 1963. *Washington: Capital City, 1879–1950.* Princeton: Princeton University Press.

Hamilton, Martha M. and Thomas Grubisich. 1980. "Tysons Corner: Crossroads of Fortune." *Washington Post,* July 13.

Henry, Shannon. 2002. The *Dinner Club: How the Masters of the Internet Universe Rode the Rise* and *Fall of the Greatest Boom in History.* New York: Free Press.

Hughes, Agatha C. and Thomas P. Hughes. 2000. *Systems, Experts, and Computers: The Systems Approach to Management and Engineering, World War II and After.* Cambridge, Mass.: MIT Press.

Jackson, Kenneth T. 1985. *Crabgrass Frontier: The Suburbanization of the United States.* New York: Oxford.

Johnson, Stephen B. 1997. "Three Approaches to Big Technology: Operations Research, Systems Engineering, and Project Management." *Technology and Culture* 38, no. 4: 891–919.

Kargon, Robert, Stuart W. Leslie, and Erica Schoenberger. 1992. "Far Beyond Big Science: Science Regions and the Organization of Research and Development." In *Big Science: The Growth of Large-Scale Research*, ed. Peter Galison and Bruce Hevley, 334–354. Stanford: Stanford University Press.

Korr, Jeremy L. 2002. "Washington's Main Street: Consensus and Conflict on the Capital Beltway, 1952–2001." Dissertation, University of Maryland, 2002.

Knowles, Scott, and Stuart W. Leslie. 2001. " 'Industrial Versailles': Eero Saarinen's Corporate Campuses for GM, IBM, and AT&T." *Isis* 92, no. 1: 1–33.

Leslie, Stuart W. 1993. *The Cold War and American Science.* New York: Columbia University Press.

Lewis, Tom. 1997. *Divided Highways: Building the Interstate Highways, Transforming American Life.* New York: Viking Penguin.

Light, Jennifer S. 2003. *From Warfare to Welfare: Defense Intellectuals and Urban Problems in Cold War America.* Baltimore: Johns Hopkins University Press.

Markusen, Ann, Peter Hall, Scott Campbell, and Sabina Deitrick. 1991. *The Rise of the Gunbelt.* New York: Oxford.

Mastran, Shelly Smith. 1988. "The Evolution of Suburban Nucleations: Land Investment Activity in Fairfax County, Virginia, 1958–1977." Phd Dissertation, University of Maryland.

MIT Operations Research Center. 1959. *Notes on Operations Research.* Cambridge: MIT Press.

Merriken, John E. 1987. *Old Dominion Trolley, Too: A History of the Mount Vernon Line.* Dallas: L.O. King Jr.

Morse, Philip. 1977. *In at the Beginnings: A Physicist's Life*. Cambridge: MIT Press.

Mueller, Milton L. 2002. *Ruling the Root: Internet Governance and the Taming of Cyberspace*. Cambridge, MA: MIT Press.

Needell, Allan A. 2000. *Science, Cold War, and the American State: Lloyd V. Berkner and the Balance of Professional Ideals*. Amsterdam: Harwood Academic Publishers.

Netherton, Nan and Ross Netherton. 1992. *Fairfax County: A Contemporary Portrait*. Virginia Beach, VA: Donning.

Netherton, Nan, Donald Sweig, Janice Artemel, Patricia Hickin, and Patrick Reed. 1978. *Fairfax County: A History*. Fairfax, VA: Fairfax County Board of Supervisors.

Norberg, Arthur, and Judy E. O'Neill. 1996. *Transforming Computer Technology: Information Processing for the Pentagon, 1962–1986*. Baltimore: Johns Hopkins University Press.

Peters, Terry Spelman. 1974. *The Politics and Administration of Land Use Control: The Case of Fairfax County, Virginia*. Lexington, MA: D.C. Heath.

Plotz, David. "[Tysons Corner, Va]." *New York Times*, June 19, 2002.

Ponturo, John. 1979. "Analytical Support for the Joint Chiefs of Staff: The WSEG Experience, 1948–1976." Institute for Defense Analyses Study S-507, Arlington, VA.

Rau, Erik. 2000. "The Adoption of Operations Research in the United States During World War II." In, *Systems, Experts and Computers: The Systems Approach in Management and Engineering, World War II and After*, ed. Agatha C. and Thomas P. Hughes (Cambridge, MA: MIT Press), 57–92.

Reily, Philip Key. 1999. *The Rocket Scientists: Achievement in Science, Technology, and Industry at Atlantic Research Corporation*. New York: Vantage Press.

Reston Web Online Community, "A History of Reston, Virginia," www.reston web.com/community/history_rest.html (accessed July 1999).

Roland, Alex. *The Military-Industrial Complex*. 2001. Washington, D.C.: Society for the History of Technology/American Historical Association.

Roland, Alex and Philip Shiman. 2002. *Strategic Computing: DARPA and the Quest for Machine Intelligence, 1983–1993*. Cambridge: MIT Press.

Rose, Mark. 1990. *Interstate: Express Highway Politics, 1939–1989*, revised edition. Knoxville: University of Tennessee Press.

Salus, Peter H. 1995. *Casting the Net: From ARPANET to INTERNET and Beyond.* Reading, MA: Addison-Wesley.

Shear, Michael D. 2003. "Hoping to 'Get Things Done' Again," *Washington Post,* June 5.

Singer, P. W. 2003. *Corporate Warriors: The Rise of the Privatized Military Industry.* Ithaca, NY: Cornell University Press.

Smith, Leef. 2002. "Despite VA Cuts, George Mason University Sets an Ambitious Course." *Washington Post,* April 21.

Stuntz, Connie Pendleton, J. Harry Shannon, and Mayo Sturdevant Stuntz. 1990. *This Was Tysons Corner: Facts and Photos.* Vienna, VA: self-published.

Swisher, Kara. 1998. *AOL.COM: How Steve Case Beat Bill Gates, Nailed the Netheads, and Made Millions in the War for the Web.* New York: Random House.

Tennenbaum, Robert, ed. 1996. *Creating a New City: Columbia, Maryland.* Columbia, MD: Perry Publishing.

Twelve Southerners. 1930. *I'll Take My Stand: The South and the Agrarian Tradition.* New York: Harper and Brothers.

U.S. Bureau of the Budget. 1966. Circular A-76. Washington, D.C.: GPO.

U.S. Congress Office of Technology Assessment. 1995. "A History of the Department of Defense Federally Funded Research and Development Centers." OTA report no. OTA-BP-ISS-157. Washington, D.C.: GPO.

U.S. House of Representatives, Subcommittee of the Committee on Government Operations. 1963. *Systems Development and Management.* Washington, D.C.: Government Printing Office, 1963.

U.S. Naval Ordnance Laboratory and Naval Surface Warfare Center. 2000. *Legacy of the White Oak Laboratory.* Dahlgren, VA: U.S. Navy.

Waldrop, M. Mitchell. 2001. *The Dream Machine: J. C. R. Licklider and the Revolution That Made Computing Personal.* New York: Viking.

Williams, Ames W. 1984. *The Washington and Old Dominion Railroad, 1847–1968.* Alexandria, VA: Meridian Sun Press.

Williamson, Mary Lou, ed. 1987. *Greenbelt: History of a New Town, 1937–1987.* Norfolk, VA: Donning.

Zachary, G. Pascal. 1997. *Endless Frontier: Vannevar Bush, Engineer of the American Century.* New York: Free Press.

NOTES

CHAPTER ONE

1. Roger Stough, " 'Twas Uncle Sam that Spurred Beltway Boom," *Washington Technology*, http://www.washingtontechnology.com/print/15_2/1241-1.html, March 5, 2004, accessed May 25, 2007.

2. The term was apparently coined by the Northern Virginia Technology Council. Other names have also been proposed, many of which were variations of "Silicon Valley," in an attempt to show that this region was an East Coast counterpart to what was happening in California. None have taken hold, although both "Tysons Corner" and "Dulles Corridor" are probably better descriptors, and I will use them throughout the text.

3. Mark Leibovitch, "At the Height of a Joy Ride, MicroStrategy Dives," *Washington Post*, January 7, 2002, pp. A1, A10. The article noted that Microstrategy's founder, Michael Saylor, lost $65 billion in net worth that day.

4. After WorldCom purchased MCI, its headquarters was located in Mississippi, and after the merger of AOL with Time Warner, its headquarters was in New York City. But both firms retained large campuses in Ashburn, and both AOL and MCI retained their identity as divisions.

5. One additional person died later as a result of injuries sustained during the attack. In addition, the five hijackers on the plane were killed. Some of those killed in the Pentagon were employees of contractors who will be discussed later in this study.

6. For example, David Brooks writes, "If you asked a Democratic lawyer to move from a $750,000 house in Bethesda, Maryland, to a $750,000 house in Great Falls, Virginia, she'd look at you as if you had just asked her to buy a pickup truck with a gun rack and to shove chewing tobacco in her kid's mouth." "People Like Us," *Atlantic Monthly*, September 2003, 29–32.

7. An interesting historical note: President Lincoln's Emancipation Proclamation of January 1, 1863, declared that slaves in the states in rebellion were free. Thus it freed slaves in Virginia but not in Maryland, which was loyal to the Union.

8. Nan Netherton et al., *Fairfax County, Virginia: A History*, (Fairfax, VA: Farifax County Board of Supervisors, 1978), Chapter 6.

9. Ames W. Williams, *The Washington and Old Dominion Railroad, 1847–1968*, (Alexandria, VA: Meridian Sun Press, 1984), 8, 24.

10. A description of Lowe's experiments, including a pair of binoculars owned by him and a facsimile of a note from President Lincoln to General Winfield Scott in support of Lowe, are on display at the Smithsonian's National Air and Space Museum, "Looking at Earth" gallery. See also the Web site http://www.visitfairfax.org, accessed August 2005.

11. Ibid., 326.

12. Some historical accounts call this the Battle of Chantilly, but the town of Chantilly was at that time located several miles to the west of the actual site, making the use of that name misleading. Confederates often named battles after the town near where they occurred (for example, Manassas), while the Unionists named them after a nearby geographical feature (such as Bull Run).

13. In the summer of 1864 Confederate General Jubal Early attacked the federal city from the Maryland side, advancing as far as Fort Stevens, about five miles north of the White House.

14. Some estimates put the number of dead at Antietam at 3,650, with many more missing and thousands wounded. The September 11, 2001 attacks killed on the order of 3,000 Americans at four sites in New York, Pennsylvania, and Virginia.

15. A small plot of land has been preserved, on which are two modest stone monuments commemorating the deaths of Union Generals Isaac Stevens and Philip Kearney. The rest of the battlefield has been developed. A local history group has researched this battle and placed a detailed account on its Web site, http://www.espd.com/oxhill/index.htm (accessed August 2003).

16. For the notion of "Lost Cause," see comments by the developer John "Til" Hazel, made to Joel Garreau in Garreau's *Edge City* (New York: Doubleday, 1991), 359.

17. Professor J.A.N. Lee of Virginia Tech's Computer Science department related his story in the *IEEE Annals of the History of Computing* 22, no. 1 (2000): 77–78. In the

course of researching this study, I found only vestigial traces of that attitude, and I gratefully acknowledge the assistance of local librarians and archivists who no longer feel that history stopped in 1865.

18. Connie Pendleton Stuntz, J. Harry Shannon, and Mayo Sturdevant Stuntz, *This Was Tysons Corner: Facts and Photos* (Vienna, VA, 1990), Chapter 1.

19. Ibid., 8–9.

20. Ibid., 61.

21. The Dulles Access Road and Dulles Toll Road are colinear but are in fact separate roads.

22. Williams, *Washington & Old Dominion Railroad*, pp. 93–96, 107–110, 131–132. The "Hampshire" of the original name is now Mineral County, West Virginia. The W&OD rail-trail has over one million visitors a year, possibly the highest of any such trail in the country.

23. Netherton, et al., *Fairfax County* (1978) 380–381.

24. The Chain Bridge got its name from a bridge built in 1805, the third built at that location, where the roadway was suspended by chains. By the time of the Civil War the span was of a more conventional design.

25. Netherton et al., *Fairfax County* (1978), 462–463.

26. Williams, *Washington & Old Dominion Railroad,* 44 and 71.

27. Netherton et al., *Fairfax County* (1978), 598–599.

28. The line restored passenger service during World War II, terminating finally in 1951. Freight service continued until 1968. Williams, *Washington and Old Dominion Railroad,* 94–95.

29. George W. Hilton and John F. Due, *The Electric Interurban Railways in America* (Stanford: Stanford University Press, 1960), 328–329.

30. Constance McLaughlin Green, *Washington: Capital City, 1879–1950* (Princeton University Press, 1963), 172–173.

31. Gregg B.. Walker, *The Military Industrial Complex* (New York: Peter Lang) 1992.

32. Fairfax County Chamber of Commerce, Seventy-Fifth Anniversary pamphlet, 2000, 25.

33. One can find grits at many restaurants in Loudon County, but the area around Dulles Airport is increasingly the home of chain restaurants whose menus are identical across the country and typically do not feature that Southern specialty.

34. Robert Kargon, Stuart Leslie, and Erica Schoenberger, "Far Beyond Big Science: Science Regions and the Organization of Research and Development," in *Big Science: The Growth of Large-scale Research*, eds. Peter Galison and Bruce Hevly (Stanford: Stanford University Press, 1992), 334–354.

35. Ibid., 339.

36. Noyce is quoted in an interview with Dirk Hanson in *The New Alchemists: Silicon Valley and the Microelectronics Revolution* (Boston: Little Brown, 1982), 93.

37. See, for example, Sam Sifton, *A Field Guide to the YETTIE: America's Young, Entrepreneurial Technocrats* (New York: Hyperion, 2000). Also see Christine A. Finn, *Artifacts: An Archaeologist's Year in Silicon Valley* (Cambridge: MIT Press, 2001).

38. The data were compiled from a number of sources and published in the *Washington Post*. Steven Pearlstein and Neil Irwin, "Signs of Resilience Amid Region's Wreckage," *Washington Post*, May 20, 2001, pp. A1, A16–A17. The definition of "technology employment" included professional services and consulting, which characterizes much of what goes on in northern Virginia, as well as engineering and manufacturing.

39. The figures are taken from several articles published in the *Washington Post*, including an article on Internet firms by Peter Behr, "First in WAR, Peace . . . and Internet - Tech?" *Washington Post*, January 13, 1997, Business Section, pp. 5–6. Later chapters will provide a more precise definition of the nature of the businesses, the percentage in each suburb, and how those changed during the Internet bubble over the various cycles of the economy.

40. Fairfax County Chamber of Commerce, Seventy-Fifth Anniversary pamphlet, 2000, 24.

41. This is literally true of the Fairfax Towne Center shopping center, which was built on the site of the Battle of Chantilly (Ox Hill).

42. Netherton, et al., *Fairfax County* (1978), 334.

CHAPTER TWO

1. Terence O'Hara, Editor, "Post 200," supplement to the *Washington Post*, April 28, 2003, p. 69.

2. The essay was eventually published as "As We May Think" in the *Atlantic Monthly* in 1945. For a discussion of Bush's early work on information retrieval, see Colin Burke, *Information, and Secrecy: Vannevar Bush, Ultra, and the Other MEMEX* (Metuchen, N.J.: Scarecrow Press, 1994). His role as an inspiration for later developers of the Internet and World Wide Web is documented in James Gillies and Robert Cailliau, *How the Web was Born: The Story of the World Wide Web* (New York: Oxford University Press, 2000), chapter three.

3. This paragraph and following descriptions of Bush's stay in Washington are taken largely from G. Pascal Zachary, *Endless Frontier: Vannevar Bush, Engineer of the American Century* (New York: Free Press, 1997).

4. Ibid., 83–85; also in Alex Roland, *Model Research: The National Advisory Committee for Aeronautics, 1915–1958* (Washington, D.C.: NASA, 1985), vol. 1. See also James Phinney Baxter III, *Scientists against Time* (New York: Little Brown, 1946), Chapter 1.

5. Draft of the enabling legislation, quoted in Baxter, *Scientists against Time*, 14. Zachary, *Endless Frontier*, 112, quotes a slightly different wording. The proposal specifically excluded aeronautical research, which the NACA was performing for the military (although not as well as Bush desired).

6. Zachary, *Endless Frontier*, 108–109; Baxter, *Scientists Against Time*, 16. Delano was a well-known figure in Washington and a man of many talents and achievements. He is remembered in Washington as the chair of the Washington Metropolitan Regional Authority—the predecessor of today's Planning Commission—and, as such, one who planned the orderly growth of the city before the war. See Constance McLaughlin Green, *Washington: Capital City, 1879–1950* (Princeton, N.J., 1963), 288–289.

7. In 1941 the NDRC was effectively subsumed under a new organization, the Office of Scientific Research and Development, which was funded by Congressional appropriations. Bush became its director, and its headquarters remained at the Carnegie Institution's building on 16th and P Streets, NW.

8. Baxter, *Scientists against Time*, appendix C, lists 25 major "non-industrial" and "industrial" contracts let by the OSRD as of June 30, 1945.

9. The phrase is the subtitle of Zachary's book, *Endless Frontier*.

10. Zachary, *Endless Frontier*, 115, quotes political scientist Donald K. Price as the source of the phrase.

11. Erik P. Rau, "The Adoption of Operations Research in the United States During World War II," in *Systems, Experts, and Computers: The Systems Approach in Management and Engineering, World War II and After*, ed. Agatha C. and Thomas P. Hughes (Cambridge, MA: MIT Press, 2000), 57–92.

12. Ibid.

13. Baxter, *Scientists against Time*, chapter 9.

14. Philip Morse, *In at the Beginnings: A Physicist's Life* (Cambridge: MIT Press, 1977), 174.

15. Martin Collins, *Cold War Laboratory: RAND, the Air Force, and the American State, 1945–1950* (Washington, D.C.: Smithsonian Institution Press 2002), 53. Collins calls RAND "a loose acronym of "research and development." Some have stated that it stood for "research and no development" but there is no evidence to support that.

16. See, for example, Paul Dickson, *Think Tanks* (New York: Ballantine Books, 1971). The popular press played up the notion of "long-haired" (although many wore were crew cuts) scientists in their casual shirts, smoking pipes and staring out over the Pacific Ocean as part of their daily work.

17. Collins, *Cold War Laboratory*, 48–49.

18. U.S. Congress Office of Technology Assessment, "A History of the Department of Defense Federally Funded Research and Development Centers," Washington, D.C.: GPO, OTA report no. OTA-BP-ISS-157, June 1995, 14–15.

19. Ibid., 23.

20. Ibid., 16.

21. Ibid., 20–22.

22. Air Ministry [UK], *The Origins and Development of Operational Research in the Royal Air Force*, Air Publication 3368 (London: Her Majesty's Stationery office, 1963), xviii.

23. See for example MIT Operations Research Center, *Notes on Operations Research* (Cambridge: MIT Press, 1959). Current descriptions of these techniques are found in Anthony Ralston and Edwin D. Reilly, eds., *Encyclopedia of Computer Science,* third edition (New York: Van Nostrand Reinhold, 1993).

24. Janet Abbate, *Inventing the Internet* (Cambridge: MIT Press, 1999).

25. Charles and Ray Eames, *A Computer Perspective* (Cambridge: Harvard University Press, 1973), 112–113.

26. Jean E. Sammet, *Programming Languages: History and Fundamentals* (Englewood Cliffs, N.J.: Prentice Hall, 1969), Chapter 9; also Thomas J. Bergin and Richard Gibson, eds., *History of Programming Languages*, vol. 2. (Reading; Addison-Wesley 1996), Chapter 8.

27. Harry M. Markowitz, oral history conducted by Jeffrey Yost of the Charles Babbage Institute, March 18 2002. Minneapolis MN; Charles Babbage Institute Center for the History of Information Technology; Arch.

28. Stanislaw M. Ulam, *Adventures of a Mathematician* (New York: Scribner, 1976), Chapter 10.

29. The theory of random numbers takes up a major portion of the classic textbook on computer programming by Donald Knuth, *The Art of Computer Programming, Vol. 2: Seminumerical Algorithms* (Reading; MA: Addison-Wesley, 1969), chapter 3. Knuth defines the Monte Carlo method in much more general terms: " . . . a general term used to describe any algorithm that employs random numbers." (p 2).

30. The book may not be considered a classic along with *Moby Dick*, but it is so considered among mathematicians. A lot of folklore has grown up around its publication, including the story that the New York Public Library classified it under "Psychology." Most research libraries have a copy, and recently it has been made available online from the RAND Web site, www.rand.org.

31. Robert Dorfman, "The Discovery of Linear Programming," *Annals of the History of Computing* 6 (July 1984): 283–295.

32. As with any discussion of "firsts," there is a great deal of controversy over these claims. For further discussion, see Paul Ceruzzi, *Beyond the Limits: Flight Enters the Computer Age* (Cambridge: MIT Press, 1989) and Ceruzzi, *A History of Modern Computing* (Cambridge: MIT Press, 1998).

33. This assertion depends on the definition of "cabinet-level" federal agency. One could argue that the CIA is such an agency. Unlike the Pentagon, the CIA headquarters in Langley, Virginia does not lie within the original ten-square-mile parcel ceded from Virginia and Maryland to form the seat of government.

34. Alfred Goldberg, *The Pentagon: the First Fifty Years* (Washington, D.C.: Historical Office, office of the Secretary of Defense, 1992), 34. Later on some of this land,

still under federal control, was transferred to Fort Myer and Arlington National Cemetery.

35. In the 1940s and 1950s this ruling made Virginia laws pertaining to racial segregation not applicable to those who worked inside the building.

36. Goldberg, *The Pentagon*, 6–7.

37. The building is now part of the offices of the State Department.

38. David Brinkley, *Washington Goes to War* (New York: Ballantine, 1988), xii.

39. Felix Gillette, "Sleepy Hollow: Tales of the District's Drowsiness are Getting Tired," *Washington City Paper*, July 14–24, 2003.

40. Goldberg, *The Pentagon*, 3–22.

41. Steve Vogel, "The Battle of Arlington: How the Pentagon Got Built," *Washington Post*, April 26, 1999.

42. Goldberg, *The Pentagon*, 45, 62.

43. Ibid., Part I.

44. Ibid., 28.

45. JRDB Charter, quoted in Allan Needell, *Science, Cold War, and the American State* (Amsterdam: Harwood Academic, 2000), 111. The War Department at that time included the Army and the Army Air Corps.

46. Ibid., 111–120.

47. John Ponturo, "Analytical Support for the Joint Chiefs of Staff: The WSEG Experience, 1948–1976," IDA Study S-507 (Arlington: Institute for Defense Analyses, July 1979) chapter 1.

48. Ibid., 2.

49. Ibid.; also Weapons Systems Evaluation Group, "Handbook." Record Group 330, National Archives and Records Administration.

50. Ponturo, "Analytical Support," 43, 48.

51. Vannevar Bush to Karl T. Compton, September 30, 1949. Record Group 330, National Archives and Records Administration.

52. Ponturo, "Analytical Support," 51–56.

53. The results, published as a ten-volume report in February 1950, addressed only the first question, the delivery of atomic bombs to targets within the Soviet Union. The details are beyond the scope of this study.

54. U.S. Office of Technology Assessment, "A History of Department of Defense Federally-Funded Research and Development Centers," 1995, 26–27. Also Institute for Defense Analyses, "In Profile," brochure dated 1997. IDA also established a Communications Research Division in Princeton, New Jersey, which later was the target of anti-Vietnam war protests.

55. The Office of Technology study cited in note 54 is one such study and cites many others, as well as Congressional testimony against FFRDC.s that appears from time to time. Criticism from the private sector is best found in the publications and Web site of the Professional Services Council, a trade association initially located in Tysons Corner.

56. Ponturo, "Analytical Support," chapter 4.

57. Ibid., 158, 170–171.

58. In 1988 the distillery moved to Spotsylvania County, as Reston was undergoing a tremendous growth spurt. It remains Virginia's last whiskey distillery. The Bowman web site gives more information: www.bowmanco.com.

59. Robert Louis Benson and Michael Warner, eds., *VENONA: Soviet Espionage and the American Response, 1939–1957* (Washington, DC: National Security Agency, Central Intelligence Agency, 1996).

60. Kathleen A. Tobin, "The Reduction of Urban Vulnerability: Revisiting 1950s American Suburbanization as Civil Defence," *Cold War History* 2, no. 2 (January 2002): 1–32.

61. U.S. Department of Energy, Office of Management and Administration, History Division, "Germantown Site History," undated brochure, circa 1997.

62. Tobin, "Reduction of Urban Vulnerability," p. 17; also Thomas Lassman, "Government Science in Postwar America: Henry Wallace, Edward London, and the Militarization of the National Bureau of Standards, 1945: *Isis* 96 (2005); 25–51.

63. James Bamford, *The Puzzle Palace* (New York: Penguin, 1982), 82–87.

64. Baxter, *Scientists against Time*, Chapter 15.

65. William B. Anspacher, Betty H. Gay, Donald E. Marlowe, Paul B. Morgan, and Samuel J. Raff, *The Legacy of the White Oak Laboratory* (Dahlgren, VA: Naval Surface Warfare Center, 2000). Information on the NIH and the Bureau of Standards was found at the NIH, NIST, and Gaithersburg, Maryland Web sites (www.nih.gov, www. nist. gov, and www.gaithersburgmd.gov.respectively).

66. Goddard Space Flight Center, *The Early Years: Goddard Space Flight Center* (Washington, D.C.: National Aeronautics and Space Administration, 1962), chapter 3.

67. See, for example, Maryann P. Feldman and Johanna L. Francis, "Homegrown Solutions: Fostering Cluster Formation," *Economic Development Quarterly* 18, no. 2 (2004): 127–137; also Margaret Pugh O'Mara, *Cities of Knowledge: Cold War Science and the Search for the Next Silicon Valley* (Princeton: Princeton University Press, 2005).

68. Numerous Web sites give details about these sites. For Mount Weather, see "FEMA Wants to Buy Land to Create New Entrance," *Winchester Star*, December 23, 2002.

69. It might be assumed that the President and perhaps senior military planners might be killed or incapacitated by an attack on Washington, but there were other contingency plans to ensure the survival of some authority.

70. These were at Annapolis and Cheltenham, MD (Navy), Woodbridge, VA and La Plata, MD (Army), and Brandywine and Davidson, MD (Air Force).

71. "Justification of the Radio Relay System from Alternate Joint Communications Center to the Washington Area," 1951, Record Group 111, National Archives and Records Administration. Relay towers were also built at Damascus and Silver Hill, Maryland.

CHAPTER THREE

1. Mary Lou Williamson, ed. *Greenbelt: History of a New Town, 1937–1987* (Norfolk, VA: Donning, 1987), Chapter 2.

2. Lewis Mumford, *The City in History* (New York: Harcourt, Brace, and World, 1961).

3. Nancy O. Phillips, ed., *Greenbelt: History of a New Town, 1937–1987* (Norfolk, Virginia: Donning, 1987).

4. Ibid., pp. 153–154.

5. Ibid., 154–155.

6. Ibid., 155, 161.

7. Jeremy L. Korr, "Washington's Main Street: Consensus and Conflict on the Capital Beltway, 1952–2001" (PhD Dissertation, University of Maryland, 2002) 113.

8. Ibid., 92.

9. There was a bewildering array of plans for roads in all directions; see the Web site "Washington, D.C. Interstates and Freeways," www.roadstothefuture.com/ D.C._interstate_Fwy.html, accessed March 2003.

10. Korr, "Washington's Main Street," 82.

11. Both quotations from Korr, "Washington's Main Street," 104.

12. Tom Lewis, *Divided Highways* (New York: Viking Penguin, 1997), 114–117.

13. Mark H. Rose, *Interstate: Express Highway Politics, 1939–1989*, revised edition (Knoxville, TN: University of Tennessee Press, 1990), 73.

14. Lewis, *Divided Highways*, 90.

15. Eisenhower quoted in Kenneth T. Jackson, *Crabgrass Frontier: The Suburbanization of the United States* (New York: Oxford, 1985), 249.

16. Korr, "Washington's Main Street," 107.

17. T. A. Heppenheimer, "The Rise of the Interstates", *American Heritage of Invention and Technology* 7, no. 2 (1991): 8–18.

18. Korr, "Washington's Main Street," 129–130; also Smithsonian Institution, National Museum of Natural History, Department of Botany, "About Plummers Island," http://persoon.si.edu/dcflora/D.C.Plummers/, (accessed November 2003).

19. George W. Hilton and John F. Due, *The Electric Interurban Railway in America* (Stanford: Stanford University Press, 1960), 328–329; also LeRoy O. King, Jr., *100 Years of Capital Traction* (Dallas, TX: Taylor Publishing, 1972), 181.

20. Donald L. Hymes, "Beltway Routing Will Spare Trees," *Washington Post*, July 11, 1963.

21. Maggie Locke, "Pimmit Hills: Coming of Age as a Neighborhood in Fairfax County," *Washington Post,* August 18, 1977.

22. Joel Garreau, *Edge City* (New York: Doubleday, 1991), 358.

23. Quoted in Korr, "Washington's Main Street," 134.

24. Ibid., chapter 4.

25. "Fairfax County Beltway Bandit Gets 30 Years," *Washington Post*, August 20, 1963, and "Crime in the Suburbs," ibid., January 27, 1968.

26. The Beltway Bandits Flyball team in Fairfax County is a club of dog-owners whose dogs are trained to chase balls over a specified course for points. The piece by Frank Zappa appears on his 1986 "Jazz from Hell" album. In an e-mail to the author in October 2000, the coach of the Flyball team stated that he was an employee of a local systems integration firm, which he declined to identify. He also stated that he never got any negative reaction to the use of that term.

27. Sources for this section have been taken primarily from the National Air and Space Museum Washington, D. C., archives Division, Tech Files, "Airports, Virginia, Dulles, 1950s."

28. Eugene Scheel, "Dulles Airport Has Its Roots in Rural Black Community of Willard," *Washington Post*, November 17, 2002.

29. National Air and Space Museum, Washington, D. C., Archives Division, Tech Files, "Airports, Virginia, Dulles, 1950s." Some of the land for the east-west runway was apparently acquired before it was decided to shift that alignment, which leads to a slight bump of the property on the west side.

30. Ames W. Williams, The *Washington and Old Dominion Railroad, 1847–1968* (Alexandria, VA: Meridian Sun Press, 1984): 109.

31. According to informal accounts, the boundary for commercial development was to have been the Dulles Access Road, but in later years the systems integration firm PRC built its headquarters off Lewinsville Road, just to its northeast.

32. Helen Dewar, "Future Looks Vast for Tysons Corner," *Washington Post*, November 30, 1963. Also see Connie Pendleton Stuntz, J. Harry Shannon, and Mayo Stuntz, *This Was Tysons Corner: Facts and Photos* (Vienna, VA: Self-published, 1990): 62–63.

33. "10th Anniversary, Dulles International Airport," *Washington Sunday Star* special supplement, November 17, 1972.

34. Ibid., v-7. The Interceptor cost $28 million and was paid for by a separate appropriation from Congress.

35. Nan Netherton, et al, *Fairfax County, Virginia: A History* (Fairfax, VA: Fairfax County Boars of Supervisors, 1978), 465.

36. Ibid. Also see Reston Webonline Community, "A History of Reston, Virginia," www.restonweb.com/community/history_rest.html (accessed July 1999).

37. Robert E. Simon, video interview with the Fairfax County Economic Development Authority, in "Fairfax County: Creating a Modern Economy," collection of video interviews published in 2003 by the Fairfax County Economic Development Authority, Vienna, VA.

38. Frank Lloyd Wright, *The Living City* (New York: Horizon Press, 1958), a compilation and revision of ideas originally published in 1932 as *The Disappearing City*, and in 1945 as *When Democracy Builds*. The scale model of Broadacre City was seen by the author at Taliesin, in Wisconsin.

39. Lewis Mumford, *Technics and Civilization* (New York: Harcourt, 1934). Lewis Mumford's prewar writings touted the advantages of the automobile and airplane, but his later writings reversed that opinion, as he witnessed and chronicled the effect of the automobile on cities, and the airplane's use as a weapon in war.

40. Wright, *The Living City*, 1958, front fold-out. Reston was served by the W&OD railroad and there were grade crossings, but traffic by the mid-1960s was negligible, and the road was shut down in 1968.

41. Ibid. Reston is not an incorporated town, unlike neighboring Vienna or Falls Church.

42. Reston Website, "A History of Reston, Virginia."

43. An almost identical plan was being developed for the new town of Columbia, Maryland, halfway between Washington and Baltimore. Columbia had a few interesting differences, however, which space does not permit a discussion of here. The main difference is that it was a stated a goal for Columbia (one not explicitly mentioned for Reston) to be a racially integrated community, as well as a community of working and white-collar classes (as was Reston). See Robert Tennenbaum, ed., *Creating a New City: Columbia, Maryland* (Columbia: Perry Publishing, 1996).

44. By contrast, the developers of Columbia built a rail spur to the nearby mainline of the B&O Railroad to service the manufacturing facilities they planned for the town. Factories did indeed sprout up in Columbia, but the rail spur has never seen

much use. Like Reston, Columbia owes much of its success to its location near a major airport, but neither Columbia nor Reston has passenger rail service.

45. James Cleveland (president of Reston Land), interview with the author, April 6, 2000.

46. Wright was a champion of the high-rise office building, but he envisioned them standing alone, surrounded by generous amounts of green space.

CHAPTER FOUR

1. U.S. Office of Technology Assessment (1995), A History of the Department of Defense Federally Funded Research and Development Centers, 20–22.

2. Joel Garreau, *Edge City: Life on the New Frontier* (New York: Doubleday, 1991), part IV; also see Nan and Ross Netherton, *Fairfax County: A Contemporary Portrait* (Virginia Beach, VA: Donning, 1992), 42–43.

3. James J. Leto, CEO of Planning Research Corporation, quoted in Kathleen Day, "Making Teamwork Work: In a Shrinking Industry, PRC Discovers There's no Room for Divisiveness," *Washington Post* December 19, 1994, Business Secion, pp. 1, 12.

4. Philip Key Reily, *The Rocket Scientists: Achievement in Science, Technology, and Industry at Atlantic Research Corporation* (New York: Vantage Press, 1999).

5. G.T. Halpin (Virginia real estate developer), interview with the author, March 31, 2003. Also Robert Reed, "From 'Futuristic' to 'Bizarre' as the Years Pass," *Alexandria Gazette*, August 28, 1988.

6. Claude Baum, *The System Builders: The Story of SDC*. (Santa Monica, CA: System Development Corporation, 1981): photo after 86.

7. *40th Year RAND* (Santa Monica: RAND Corporation, 1988): 16–17.

8. Reily, *The Rocket Scientists*, pp. 132–133 color plate following p. 112.

9. Shelly Smith Mastran, "The Evolution of Suburban Nucleations: Land Investment Activity in Fairfax County, Virginia, 1958–1977". (PhD Dissertation, University of Maryland, 1988).

10. Ibid., chapter III.

11. Ibid., 110.

12. Ibid., 75.

13. Martha M. Hamilton and Thomas Grubisich, "Tysons Corner: Crossroads of Fortune," *Washington Post*, July 13, 1980. Also see John Brooks, *The Go-Go Years* (New York: Dutton, 1984).

14. Hamilton and Grubisich, "Tysons Corner."

15. Ibid.

16. Mastran, "Evolution of Suburban Nucleations," 90–91.

17. Hamilton and Grubisich, "Tysons Corner."

18. As of this writing, for example, many regard the Thomas Jefferson High School of Science and Technology, located in Alexandria, as the region's best. And the shopping malls at Tysons Corner are likewise regarded as the region's most upscale. Neither of these were present in the mid-1960s.

19. An extensive discussion of Virginia highways, built as well as unbuilt, was found at the Web site www.roadstothefuture.com/roadsnova, accessed March 2003.

20. Mastran, "Evolution of Suburban Nucleations," 66.

21. Ellen McCarthy, "Technology Center Modifies its Focus to Secure Funding," *Washington Post,* March 18, 2004.

22. Loring Wirbel, "Trappings of Empire: The Escalating Costs of Space Control," Citizens for Peace in Space, www.space4peace.net/articles/escalatingcosts.htm.

23. The CIA attack was carried out by Mir Aimal Kansi, who fired his AK-47 assault rifle into cars waiting to turn in the CIA's driveway at Chain Bridge Road, killing two and injuring three. Kansi was later captured in Pakistan, extradited to the U.S., and executed.

24. Jeffrey Richelson, *The Wizards of Langley: Inside the CIA's Directorate of Science and Technology* (Boulder, CO: Westview Press, 2001).

25. Benjamin Forgery, "Crystal-Gazing at News Headquarters," *Washington Post*, January 12, 2002.

26. Web site of United American, Inc., Commercial General Contractors. Vienna, Virginia, http://www.unitedamericaninc.com/who.html.

27. Christopher Alexander, *A Pattern Language* (New York: Oxford, 1977).

28. "One Compiler, Coming Up!" *Datamation* (May/June 1959): 15. Also see Saul Rosen, "Programming Systems and Languages: A Historical Survey," *Proceedings Spring Joint Computer Conference* 25 (1964): 1–15.

29. Baum, *The System Builders*.

30. Ibid., 73–74.

31. ANSER Annual Reports, date, Box 14, Philip Morse Papers, MIT Archives.

32. Kent C. Redmond and Thomas M. Smith, *From Whirlwind to MITRE: The R&D Story of the SAGE Air Defense System* (Cambridge, MA: MIT Press, 2000), 411–428.

33. MITRE, "History of the MITRE Corporation," www.mitre.org/about/history.html (accessed June 13, 2002).

34. Liza Bercovici, "The Study Merchants," *Washington Post*, November 12, 1978.

35. Ann Markusen, Peter Hall, Scott Campbell, and Sabina Deitrick, *The Rise of the Gunbelt: The Military Remapping of Industrial America* (New York: Oxford, 1991).

CHAPTER FIVE

1. David H. DeVorkin, *Science with a Vengeance: How the Military Created the U.S. Space Sciences After World War II* (New York: Springer, 1992), Chapter 7.

2. Braddock, Dunn & McDonald, Inc, "Organization and Professional Staff," brochure, page 3, ca. 1962. In author's possession.

3. Harry Jaffee, "The High Priest of High-Tech: Earle Williams and the Future of Fairfax County," *Regardie's*, (July 1985), 58–63.

4. Braddock, Dunn, and McDonald, "Annual Report," 1991 copy in the author, posession.

5. Earle Williams, interview with the author, June 29, 2000. Also, Fairfax County telephone directory, 1968–1969.

6. BDM International, Inc., Securities and Exchange Commission Form 10-K, 1982, page 2. Copy supplied to the author G Earle Williams.

7. Davis Dyer, *TRW: Pioneering Technology and Innovation Since 1900* (Boston: Harvard University Press, 1998).

8. H. L. Nieburg, *In the Name of Science* (Chicago: Qudrangle Books, 1966), Chapter 10.

9. Ibid.

10. Earle Williams, interview with the author, June 29, 2000. Also BDM International, Inc., Securities and Exchange Commission Form 10-K, 1982 and BDM Annual Reports, 1977–1991.

11. Allan A. Needell, *Science, Cold War and the American State: Lloyd V. Berkner and the Balance of Professional Ideals.* (Amsterdam: Harwood, 2000).

12. David A. Hounshell, "After September 11, 2001: An Essay on Opportunities and Opportunism, Institutions and Institutional Innovation, and Technologies and Technological Change," *History and Technology* 19, no. 1 (March 2003): 39–49.

13. John Ponturo, "Analytical Support for the Joint Chiefs of Staff" (Arlington, VA: IDA, 1979), 180–186.

14. Fred Kaplan, *The Wizards of Armageddon* (New York: Simon & Schuster, 1983), 252–253. Also see Ponturo, "Analytical Support for the Joint Chiefs of Staff," 198–199.

15. Ponturo, "Analytical Support for the Joint Chiefs of Staff," 198–199.

16. Ibid., 207.

17. Joseph Braddock, interview with the author, September 10, 1997.

18. John W. Finney, "Navy Missile Project Illustrates Interaction of Contractors, Consultants, and the Military," *New York Times*, July 26, 1976), p. 12.

19. Martha M. Hamilton, "Flourishing Federal Contractors Help to Fuel Region's Economy," *Washington Post*, February 17, 1980. F1, F6.

20. Joseph Braddock, interview with the author, September 10, 1997.

21. Beth Bersilli, "Solutions, Incorporated," *Washington Post,* November 10, 1977.

22. Albert Lavagnino, interview with the author, January 14, 1998. The dollar figure for the hammer was quoted by Alex Roland in *The Military-Industrial Complex* (Washington, D.C.: Society for the History of Technology/American Historical Association 2001), 17.

23. P. W. Singer, *Corporate Warriors: The Rise of the Privatized Military Industry* (Ithaca, NY: Cornell University Press, 2003); also Jennifer S. Light, *From Warfare to Welfare: Defense Intellectuals and Urban Problems in Cold War America* (Baltimore: Johns Hopkins University Press, 2003).

24. See, for example, the "Note on Sources" written by Alex Roland and Philip Shiman for their study of the Defense Advanced Research Projects Agency: *Strategic Computing: DARPA and the Quest for Machine Intelligence, 1983–1993* (Cambridge, MA: MIT Press, 2002): 397–403, where the authors lament that many records of DARPA, a federal agency, are kept in the offices of DynCorp.

25. Michael Schrage, "Why Did Venture Boom Bypass D.C.?" *Washington Post*, May 26, 1986, 1, 10.

26. James Lardner, "You Too, can be a Consultant: Practitioner Divulges Secrets in Seminars for Free-Lancers," *Washington Post*, May 13, 1978. Also see the Professional Services Council membership directories for 1990, 1991, 1992, and 1993. Also Sam Fishbein (historian, National Air and Space Museum), interview with the author, November 7, 2000.

27. Ellen McCarthy, "Government Clears CACI for Contracts," *Washington Post*, July 8, 2004. The Iraq interrogation contract that got CACI into trouble was not let by Defense but by the Interior Department, "to quickly purchase information technology services from the company."

28. Section 8(a) of the Small Business Act 15, U.S.C. 637(a); also Federal acquisition Regulation (FAR) 19.8. See http://www.sba.gov/library/cfrs/13cfr124.html, accessed October 2004.

29. The report was reprinted in U.S. House of Representatives, Subcommittee of the Committee on Government Operations, *Systems Development and Management* (Washington, D.C.: Government Printing Office, 1963).

30. Executive Office of the President, Bureau of the Budget, Circular No. A-76, March 3, 1966. It has been modified many times, and remains in force. The BoB is now the Office of Management and Budget (OMB).

31. The 1999 revision may be found at http://www.whitehouse.gov/omb/circulars/a076/a76-rev 2003.pdf, accessed June 7, 2007.

32. See, for example, the Defense Logistics Agency's extensive Web site devoted to acquisition: http://www.dla.mil/j-3/a-76/A-76Main.html.

33. The government Web site is now http://www.fedbizopps.gov; see also http://cbdnet.gpo.gov.

34. Lester Shubin (former employee of MelPar), interview with the author, October 16, 2000. Also see Arthur Norberg and Judy O'Neil, *Transforming Computer Technology: Information Processing for the Pentagon, 1962–1986* (Baltimore, Johns Hopkins University Press, 1996), and Alex Roland and Philip Shiman, "Note on Sources" *Strategic Computing*.

35. Roland and Shiman, *Strategic Computing*, 112–113.

36. The ARPANET will always remain an example of the enlightened use of federal money to stimulate a technology, but other than that major exception, ARPA's record has been mixed. See Norberg and O'Neill, *Transforming Computer Technology*.

37. Earle Williams, interview with the author, June 29, 2000.

38. This policy is still in force as of this writing, although there are exceptions.

39. Roland, *The Military-Industrial Complex*, 26–27.

40. BDM International, Annual Report, 1991. Also see Dan Briody, *The Iron Triangle: Inside the Secret World of the Carlyle Group* (New York: John Wiley, 2003), chapter 4, and Earle Williams, interview with the author, June 29, 2000.

41. Harry M. Markowitz, interview with Jeffrey Yost, March 18, 2002, transcript Oral History 333, Charles Babbage Institute.

42. National Museum of American History, Division of Computers, Information, and Society, curatorial files, box temp 212 304 349.R019.

43. Renae Merle, "Jack London's Calling: CACI's Future Depends on Increased Defense Spending," *Washington Post*, November 12, 2001. Also see Anitha Reddy, "CACI Hungers to Reach Top Tier," *Washington Post*, October 20, 2003.

44. John Toups (former CEO and chairman of PRC), interview with the author, October 26, 2000.

45. Stephen B. Johnson, "Three Approaches to Big Technology: Operations Research, Systems Engineering, and Project Management," *Technology and Culture* 38 (October 1997): 891–919.

46. For a general discussion of the different types of computers, see Paul E. Ceruzzi, *A History of Modern Computing*, Second Edition (Cambridge, MA: MIT Press, 2003), chapter 4.

47. Toups, interview with the author.

48. Fairfax County telephone directory, 1970.

49. Larry Armstrong, "Happy Fallout Down at the Nuke Lab," *Business Week* (October 7, 1996), 42; also SAIC Web site, www.saic.com.

50. The report of 1998 lists five, namely health care, telecommunications, national security, energy and environment, and transportation and logistics.

51. Steven Pearlstein, "A Beltway Bubble About to Burst?" *Washington Post*, June 20, 2003, EI also Anitha Reddy, "SAIC's New Chief Has Big Plans," *Washington Post*, February 6, 2004.

52. The Federation of American Scientists' Web page on the "Intelligence Industry" listed the addresses of these SAIC facilities as of March 2003: see www.fas.org/irp/contract/s.htm, accessed October 2004.

53. Steven Pearlstein, "A Beltway Bubble About to Burst?" *Washington Post*, June 20, 2003; also Carolyn T. Geer, "Turning Employees into Stakeholders," *Forbes*, December 1, 1997.

54. Most of this information was taken from the Dyncorp Web site www.dyncorp.com, accessed October 2004.

55. Rajiv Chandrasekaran, "DynCorp Seeks a Dose of the Health-Care Market," *Washington Post*, February 24, 1997.

56. Renae Merle, "Computer Sciences Plans to Acquire DynCorp," *Washington Post*, December 14, 2002.

57. Jennifer S. Light, *From Warfare to Welfare: Defense Intellectuals and Urban Problems in Cold War America* (Baltimore: Johns Hopkins, 2003).

58. Steve Raney and Stan Young, "Morgantown People Mover—Updated Description," TRB 2005 Reviewing Committee: Circulation and Driverless Transit (AP040), http://www.cities21.org/morgantown_TRB_111504.pdf. In 2004 the federal and local governments approved a plan to extend the Washington Metro (not a PRT system) to Tysons Corner. It is scheduled to open in 2012.

59. Jo Thomas, "Congress Questions Advantage and Cost of U.S. Consultants," *New York Times*, December 5, 1977.

CHAPTER SIX

1. Martha M. Hamilton and Thomas Grubisich, "Tysons Corner: Crossroads of Fortune," *Washington Post*, July 13, 1980.

2. Connie Stuntz, J. Harry Shannon, and Mayo Stuntz, *This was Tysons Corner: Facts and Photos* (Vienna, VA: 1990), 103–146.

3. Stuntz, Shannon, and Stuntz, 46, 53 (n. 11).

4. Megan Rosenfeld, "Tysons Corner a Model for Suburbia's Future," *Washington Post,* Feburary 20, 1977.

5. Thomas G. Morr (managing partner of the Greater Washington Initiative), interview with the author, March 23, 2003.

6. "Tysons Corner Center—New and Upcoming Stores, Summer 2000," informational brochure in author's possession.

7. Robert E. Lang, *Edgeless Cities: Exploring the Elusive Metropolis* (Washington, D.C.: Brookings, 2003), 49.

8. The author has no specific information as to the current use of the tower, but in 2004 a new geodesic dome suddenly appeared on the top of the tower, representing a technology not in use during the Cold War.

9. "High-Rise to Fall," *Washington Post*, February 4, 2002.

10. "Tysons Shopping Center Puts Dollar Squeeze on Neighbors," *Washington Post*, May 20, 1969.

11. Wolf von Eckardt, "The Beautiful Beltway Heralds Urban Mess," *Washington Post*, October 25, 1964.

12. In particular, the Occoquan River, which formed the southwestern boundary of the county, was heavily stressed by development.

13. Terry Spielman Peters, *The Politics and Administration of Land Use Control: The Case of Fairfax County, Virginia* (Lexington, MA: D.C. Heath, 1974).

14. Ibid. Also see Nan Netherton, et al. *Fairfax County, A History*, (Fairfax, VA: Fairfax County Board of Supervisor. 1978), 640. The numerous lawsuits are described in Joel Garreau, *Edge City* (New York: Doubleday, 1991), 382.

15. Michael D. Shear, "Hoping to Get Things Done Again," *Washington Post*, June 5, 2003.

16. Commonwealth of Virginia, County of Fairfax, "Committee to Study the Means of Encouraging Industrial Development in Fairfax County," Typescript in the author's possession, informally known as the "Noman Cole Report" after one of its authors.

17. Fairfax County Economic Development Authority, "Fairfax County: Creating a Modern Economy," videotaped interviews conducted with Earle Williams and John T. Hazel at Northern Virginia Community College, October 2003.

18. Ibid.; also Earle Williams, interview with the author, June 29, 2000.

19. Fairfax County Economic Development Authority, "Fairfax County: Creating a Modern Economy."

20. Williams, interview with the author, June 29, 2000.

21. Fairfax County (Virginia) Chamber of Commerce, 75th Anniversary Commemorative Publication, 2000, 24.

22. Ibid., 28.

23. The *Washington Post*, in an article on the Business Section on March 4, 2004, suggested 40 percent, but informal estimates from others are higher.

24. Shear, "Hoping to 'Get Things Done'," quoting Al Dwoskin, a local developer who was a Democrat. Also see Gerald Halpin and Katherine Maclane, interview with the author, March 31, 2003.

25. Commonwealth of Virginia, County of Fairfax, "Committ to Study the Means of Encouraging Industrial Development in, "Fairfax County," 12.

26. Much of this information is taken from the Web site of "Roads to the Future," at http://www.roadstothefuture.com, accessed March 14, 2003.

27. Reports of the Northern Virginia Transportation Alliance, copies in the author's possession.

28. Christophe Lécuyer, *Making Silicon Valley: Innovation and the Growth of High Tech, 1930–1970* (Cambridge, MA: MIT Press, 2005).

29. Mark Frankel, "George Mason's Call to Revolution," *Regardie's* (February 1984), 74–80.

30. George Johnson, interview with the author, March 4, 2004.

31. Leef Smith, "Despite VA Cuts, George Mason University Sets an Ambitious Course," *Washington Post*, April 21, 2002.

32. Ibid.

33. Ibid.

34. The University of Maryland's reputation may suffer from its having a good football program. At the same time, Johnson noted that Earle Williams liked to visit—and donate money to—his alma mater, Auburn, where he sat in the president's box at football games.

35. Shanon Henry, "George Mason Defends its High-Tech Turf," *Washington Post*, January 13, 2005.

36. Donald R. Baucom, *The Origins of SDI, 1944–1893* (Lawrence: University of Kansas Press, 1992).

37. Ibid., Chapter 8.

38. Quoted in the Council on Economic Priorities, *Star Wars: The Economic Fallout* (Cambridge, MA: Ballinger, 1987) 1; also Baucom, 130.

39. The word "blockbuster" originally meant a very large conventional bomb dropped from an airplane, and is probably of World War II origin.

40. Herb Brody, "Star Wars: Where the Money's Going," *High Technology Business* (December 1987), 22–29.

41. Ibid., 23.

42. BDM Annual Reports in the author's possession.

43. Alex Roland and Philip Shipman, *Strategic Computing: DARPA and the Quest for Machine Intelligence, 1983–1993* (Cambridge, MA: MIT Press, 2002).

44. J. Hamilton Lambert, interviewed by the Fairfax County Economic Development Authority, "Fairfax County: Creating a Modern Economy," conducted at Northern Virginia Community College, October 2003.

45. Walter Boyne, "Flying out of the Cold War," *Cosmos* (electronic-only journal), 1999 http://www.cosmos-club.org/web/journals/index.html/. Also Commission on the Future of the United States Aerospace Industry, *Final Report*, Arlington, VA: 2003, 7–1.

46. The European Union did manage to prevent the merger of two large U.S. aerospace firms, but this was an exception to the rule.

47. Commission on the Future of the United States Aerospace Industry, *Final Report*, 7–4.

48. These numbers include employment in the District and Maryland as well as Virginia Supplement to the "Post 200," *Washington Post*, Larry Liebert, editor, April 25, 2005.

49. Northrop Grumman builds the B-2 Stealth bomber. It is a major subcontractor to Boeing for parts of Boeing's commercial aircraft and for the Space Shuttle. The company also builds unmanned aerial vehicles (UAVs), which in many ways are replacing the manned aircraft of previous eras.

50. Grumman established its reputation as a builder of rugged carrier-based airplanes during World War II, but its most famous craft was the Lunar Module, the machine that took astronauts to the moon and that was not designed to fly in the atmosphere.

51. Ronald Sugar, quoted in Joseph C. Anselmo, "Where IT's At," *Aviation Week and Space Technology* (July 11, 2005), 50–51.

52. Ibid.

53. Earle Williams, interview with the author, June 29, 2000.

54. Georg Johnson, interview with the author, March 4, 2004.

CHAPTER SEVEN

1. Vice President Gore never actually claimed credit for the Internet, but that was how his statement was perceived. In an interview on CNN in March 1999 he said "During my services in the United States Congress, I took the initiative in creating the Internet."

2. Robert E. Lang, *Edgeless Cities: Exploring the Elusive Metropolis* (Washington, D.C.: Brookings Institution, 2003).

3. When the W&OD's predecessor was established before the Civil War, its goal was the coal fields of the western part of Virginia, now parts of Mineral County, West Virginia.

4. Mario Morino (entrepreneur and chairman of the Marino Institute), interview with the author, February 12, 2003.

5. Ibid. The riots in the District following the assassination of Martin Luther King in 1968 precipitated a move from District locations.

6. Joe Braddock and Albert Lavagnino, interview with the author, October 28, 1997.

7. C-E-I-R Records, Archives Division, Charles Babbage Institute, Minneapolis, MN. Also see Herbert R. J. Grosch, *Computer: Bit Slices From a Life* (Third Millennium Books, 1991), 252–253.

8. "Fairfax County: A World Computer Capital," *Fairfax Prospectus* 6, no. 2 (April 1975), 1–2.

9. Ibid., 2. See also Connie Stuntz, J. Harry Shannon, and Mayo Stuntz Sturdevant, *This Was Tysons Corner: Facts and Photos* (Vienna, VA: 1990), 121. A photograph in this book reveals how the Honeywell building stood along in a semirural setting for several years before other buildings in Westpark grew up around it.

10. The term initially was applied to ten young Army Air Force systems analysts, including Robert McNamara, who were hired by Henry Ford II in 1945 to reverse the Ford Motor Company's declining fortunes. The term later was applied to those, with similar training in mathematics or systems analysis, whom McNamara hired after becoming Secretary of Defense under President John F. Kennedy in 1961. See John A. Byrne, *The Whiz Kids: The Founding Fathers of American Business—and the Legacy They Left Us*, (New York: Doubleday, 1993); also David Halberstam, *The Best and the Brightest* (New York: Random House, 1969), 229–231.

11. Renae Merle, "Fifth Founder Quits AMS," *Washington Post*, November 25, 2002.

12. Arthur L. Norberg and Judy E. O'Neill, *Transforming Computer Technology: Information Processing for the Pentagon, 1962–1986* (Baltimore, MD: Johns Hopkins University Press, 1996), 88–89.

13. J. C. R. Licklider, quoted in M. Mitchell Waldrop, *The Dream Machine: J.C.R. Licklider and the Revolution That Made Computing Personal* (New York: Viking, 2001), 5–6.

14. Ibid., 221.

15. In 1973, as ARPA managers were working on creating the Internet protocols, *Washington Post* reporter Bob Woodward was meeting in the parking garage of the building across the street, 1401 Wilson Blvd., with a Mr. Felt, also known as "Deep Throat," and exchanging secrets about the Watergate break-in.

16. The IBM system that eventually was successful was called CMS, for Cambridge Monitoring System. It was developed in the same building in Cambridge, Massachusetts where some of MIT's work was being done, but it was not an MIT project.

17. Morino Institute, "The Genesis of the Morino Foundation," (Reston, VA: 1995), accessed electronically at www.morino.org, April 2003.

18. The best articulation of this thesis is found in Wiebe E. Bijker, Thomas P. Hughes, and Trevor Pinch, *The Social Construction of Technological Systems* (Cambridge, MA: MIT Press, 1987); also Thomas Parke Hughes, *Networks of Power: Electrification in Western Society, 1880–1930* (Baltimore: Johns Hopkins University Press, 1983).

19. Janet Abbate, *Inventing the Internet* (Cambridge, MA: MIT Press, 1999).

20. Ibid. 78–80; also ARPA Network Information Center, "Scenarios for Using the ARPANET at the International Conference on Computer Communication," typescript, (Washington, D.C., October 24–26, 1972) 62.

21. Abbate, *Inventing the Internet*, chapter 3.

22. Ibid., 80–81; also Peter H. Salus, *Casting the Net: From ARPANET to INTERNET and Beyond* (Reading, MA: Addison-Wesley, 1995), chapter 11.

23. Kara Swisher, *AOL.Com: How Steve Case Beat Bill Gates, Nailed the Netheads, and Made Millions in the War for the Web* (New York: Random House, 1998).

24. Ibid.; also West*Group, e-mail to the author, April 24, 2003.

25. Stewart Brand, ed., *Whole Earth Software Catalog* (Garden City, NY: Doubleday, 1984), 138–157.

26. Kevin Kelly, ed., *Signal: Communication Tools for the Information Age, A Whole Earth Catalog* (New York: Harmony Books, 1988), 75.

27. Gerald W. Brock, *The Telecommunications Industry: the Dynamics of Market Structure* (Cambridge, MA: Harvard, 1981).

28. Swisher, *AOL.Com*, 17–39.

29. America Online, Inc., press release, August 6, 1998, copy in author's possession.

30. Abbate, *Inventing the Internet*, chapter 4.

31. Ibid., 122; also Salus, *Casting the Net*, 29–30.

32. Brock, *The Telecommunications Industry*, 211–230.

33. Caroline E. Mayer, "MCI Rewrote the Rulebook," *Washington Post*, February 15, 2005.

34. Abbate, *Inventing the Internet*, 196.

35. Vinton Cerf, "On the Commercial Interconnections of Internet Service Providers," MCI White Paper, n.d. Accessed electronically at http://global. mci.com/, April 18, 2003.

36. Abbate, *Inventing the Internet*, 197–198.

37. Rajiv Chandrasekaran, "Making UUNet into a Very Big Deal," *Washington Post*, September 29, 1997.

38. The turmoil surrounding WorldCom's acquisition of MCI may have contributed to Sidgmore's death, at 52, in December 2003. "John W. Sidgmore" (obituary), *Washington Post*, December 12, 2003.

39. The precise location is not given out.

40. MAE is pronounced as a word. The similarity of "MAE-West" to the name of the famous actress suggests that it was the first of these Network Access Points; however, MAE-East carried much more traffic.

41. Brian Hayes, "The Infrastructure of the Information Infrastructure," *American Scientist* 85, no. 3 (May–June 1997), 214–218.

42. Shawn Young, "Why the Glut in Fiber Lines Remains Huge," *Wall Street Journal*, May 12, 2005.

43. David J. Whalen, "Communications Satellites: Making the Global Village Possible" manuscript (Washington, D.C.: NASA History Office, 2004). Also Sam Etler, private communication, June 2004.

44. Verizon, "Verizon FIOS: Quick Facts to Get you up to Speed," undated brochure placed in the author; mailbox January 2005.

45. Shawn Young, "Why the Glut . . . " estimates that "[s]ome $90 billion of fiber was laid at the height of the Internet boom."

46. Network Solutions Web site, accessed July 1999 (it is no longer active).

47. Jay P. Kesan and Rajiv C. Shah, "Fool Us Once Shame on You—Fool Us Twice Shame on Us: What We Can Learn From the Privitization of the Internet Backbone Network and Domain Name System," *Washington University Law Quarterly* 79 (2001): 89–219.

48. In the summer of 2005 a four-paragraph statement on a Department of Commerce Web site stated in an official sense that Commerce had the authority to assign names: http://www.ntia.gov, accessed July 1, 2005.

49. Milton L. Mueller, *Ruling the Root: Internet Governance and the Taming of Cyberspace* (Cambridge, MA: MIT Press, 2002).

50. Postel had the last word, however; in 1998 he hijacked the registration system for a brief period, taking it away from Network Solutions. After he made his point (and was threatened with jail), he returned it. Mueller, *Ruling the Root*, 161.

51. RFC 1400, accessed electronically April 2003, at http://www.fags.org/rfcs/rfc1400.html.

52. John Markoff, "Minor Error Throws Internet into Disarray," *New York Times* July 18, 1997. The headline is misleading; the Internet continued to run smoothly. Also Robert Lemos, "Assault on Net Servers Fails," CNET.com electronic news service, October 22, 2002.

53. David McGuire, "Getting to the Root of all E-Mail," *Washington Post*, March 29, 2002.

54. Northern Virginia Regional Park Authority, Washington & Old Dominion Railroad Regional Park brochure, 6th edition, 2004. This sign was observed in the summer of 2004.

55. Joel Brinkley, "Information Superhighway Roars Outside the Beltway," *New York Times*, October 12, 1999. Also see Catherine Yang, "Between Silicon Valley and Silicon Alley: How the Area Around Washington, D.C. Became a High-Tech Haven," *Business Week* (August 30, 1999), 168–176.

56. Mario Morino, interview with the author, February 12, 2003.

57. Earle Williams, interview with the author, June 29, 2000.

58. William McGowan, quoted in Michael Schrage, "Why did Venture Boom Bypass D.C.?" *Washington Post*, May 26, 1986.

59. Ibid.

60. http://www.nvtc.org/advertising/techtopia.php

61. Jim Clark, *Netscape Time: the Making of the Billion-Dollar Start-Up that Took on Microsoft* (New York: St. Martin's Press, 1999).

62. Mark Leibovich, "Dot-Com Halo: the Rise and Fall of Michael Saylor," *Washington Post*, January 7, 2002.

63. Shannon Henry, *The Dinner Club: How the Masters of the Internet Universe Rode the Rise and Fall of the Greatest Boom in History* (New York: Free Press, 2002).

64. The building was still active as of the spring of 2003; see its Web site http://www.11600Sunrise.com, accessed spring 2003.

65. Morino did, however, move from northern Virginia to the Cleveland, Ohio area, where he originally was from. He continued to be active in northern Virginia entrepreneurship at a reduced pace.

66. The company had a few outlet stores in states that collected no sales tax, but this was its first ordinary retail store. Martha M. Hamilton, "L.L. Bean Gets a Bigger Tent," *Washington Post*, July 29, 1999.

67. Kate Royce, "Apple's First Retail Store Seeks New Core Buyers in Washington, D.C. Area," *Washington Times*, May 16, 2001.

68. U.S. National Commission on Terrorist Attacks Upon the United States, "The 9/11 Report" (U.S. Government Printing Office, 2004), 314.

69. Thomas Hobbes, *The Leviathan*, (1660), chapter 13. Accessed electronically at http://oregonstate.edu/instruct/ph1302/texts/hobbes/leviathan-contents.html May 31, 2007.

CHAPTER EIGHT

1. Roger Stough, " 'Twas Uncle Sam that Spurred Beltway Boom," *Washington Technology*, online edition, March 5, 2004, http://www.washington technology.com/print/15-2/1241-1.html; also Mario Morino, interview with the author, February 12, 2003.

2. Daniel R. Headrick, *The Invisible Weapon: Telecommunications and International Politics, 1851–1945* (New York: Oxford University Press, 1991).

3. Ibid., chapters 10 and 11.

4. The secret was kept for many years; among the first to report it was Ronald Lewin in *Ultra Goes to War* (London: Hutchinson, 1978). Since then, other books have further elaborated on the story and modified some of Lewin's initial statements, but before 1978, very little of this was known.

5. "Cable and Wireless, A History," Web site www.cwhistory.com/history/html/CWEmpire.html, accessed December 2006.

6. Drew Lindsay, "How Big Money has Changed Washington, *Washingtonian* (November 2006), 96–101.

7. Mary Clare Fleury, "Streets of Gold," *Washingtonian* (November 2006), 102–109.

8. G. Pascal Zachary, *Endless Frontier*: Vannevar Bush, Engineer of the American Century (New York: Free Press, 1997), 115.

9. Thomas Parke Hughes, *American Genesis: A Century of Invention and Technological Enthusiasm* (New York: Viking, 1989), chapter 8.

10. Aliya Sternstein, "NASA Looks at Insourcing: New Administrator Seeks to Reclaim its Research Mission," *Federal Computer Week* (June 6, 2005), 62.

11. Gerald Connolly, WTOP Radio interview, February 9, 2005.

12. In his interview, Connolly ruled out a third alternative, shifting development to Prince Georges County, Maryland, which is close to the District and has a lot of available land. Connolly gave the high crime rate of that county as the main reason for not expanding.

13. Lisa Rein, "At Top of Ticket, Fairfax Shifts from Red to Blue," *Washington Post*, November 4, 2004.

14. Michael D. Shear, "State Coffers Benefit from Boom in N.Va.," *Washington Post*, November 10, 2004. The article quotes local analyst Steven Fuller, who notes that federal spending in the region tripled since 1990, from $10 billion to about $35 billion in 2003.

15. Jennifer S. Light, *From Warfare to Welfare: Defense Intellectuals and Urban Problems in Cold War America* (Baltimore: Johns Hopkins Press, 2003).

16. White's Ferry, located above Leesburg, also carries vehicles. It is a lovely crossing and a reminder of an age long gone in the region. It does not run when the river is high.

17. Dulles Corridor Rapid Transit Project, accessed electronically at http://www.dullestransit.com, March 2005.

18. The strongest local advocate is Jerry Kieffer, a transportation planner and consultant. See "Fundamental Gaps in National Capital Area Transportation Policy," July 12, 2000, accessed electronically at http://www.faculty.washington.edu/jbs/itrans.fairfax.htm, July 2000. For a critical view of PRT, see Sigurd Grava, *Urban Transportation Systems* (New York: McGraw-Hill, 2003), Chapter 15. Grava's book is highly recommended for an overview of the transportation challenges facing Tysons Corner and Dulles Corridor. See also Edmond W. F. Rydell, *The Transportation Renaissance: The Personal Rapid Transit Solution* (Philadelphia: eXlibris, 2000).

19. Bruno LaTour, *ARAMIS, or the Love of Technology* (Cambridge, MA: Harvard University Press, 1996).

20. John E. Merriken, *Old Dominion Trolley Too: A History of the Mount Vernon Line* (Dallas: L. O. King Jr., 1987), 120–127.

INDEX